Start Your Own
COFFEE & TEA
Store

Start Your Own
COFFEE & TEA
Store

Pfeiffer
& COMPANY

Johannesburg • London
San Diego • Sydney • Toronto

This book was developed in association with the National Business Library.

Editor: JoAnn Padgett
Production Editor: Katharine Pechtimaldjian
Editorial Assistant: Susan Rachmeler
Interior Designer: Nicola Ruskin
Page Compositor: Judy Whalen
Cover: Tom Lewis, Inc.

Library of Congress Cataloging-in-Publication Data

Start your own coffee & tea store.
 p. cm.
 "Developed in association with the National Business Library"—T.p. verso.
 Includes index.
 ISBN: 0-89384-262-1 (paper)
 1. Coffee industry—Management. 2. Tea trade—Management. 3. New business enterprises—Management. 4. Specialty stores—Management.
I. Pfeiffer & Co. II. National Buisness Library. III. Title: Start your own coffee and tea store.
HD9195.A2S73 1994
663'.93'0688—dc20 94-8601
 CIP

Printed in the United States of America.
Printing 3 4 5 6 7 8 9 10

Published by Pfeiffer & Company
8517 Production Avenue
San Diego, California 92121
United States of America

Editorial Offices:
(619) 578-5900, FAX (619) 578-2042

Orders:
(U.S.A.) (606) 647-3030
FAX (606) 647-3034

CONTENTS

PREFACE

Start Your Own Coffee & Tea Store is the result of many hours of in-depth research into the coffee and tea business. Our exclusive team of National Business Library's professional business writers and Pfeiffer & Company's professional editors has brought years of practical experience to this project, and we know that the information provided in this book will set you on the road to success.

Owning your own business can be the most exciting and rewarding venture you will ever experience. We talk to hundreds of small business owners who make comments like, "Doing something I really enjoy makes every new day a pleasure," or "If I had known I could be realizing this kind of income, rather than making my former boss wealthy, I would have started my own business years ago."

It's true! You'll never get rich working for someone else. By capitalizing on your experience, investing time and energy, and studying the proven techniques and business methods provided in this book, you will be well on the way to realizing your goals for success in your own venture. It takes courage to begin. Without a doubt, the first step is the hardest—and you have already taken it!

INTRODUCTION:
PROFITS IN YOUR COFFEE CUP

What's better than that first cup of coffee in the morning? Imagine a cup of hot custom-blended coffee made from ground whole beans from South America or Arabia. Ask any coffee lover—there's just no comparison between "store bought" and fresh coffee purchased from a specialty coffee and tea store.

And for the owners of these coffee and tea stores, there is nothing like the profits that they are making in this growing industry.

Gourmet coffees and teas used to be a treat enjoyed by a limited upper-class audience because the art of coffee roasting was a limited practice and unique beans were scarce and costly. With the increased availability of good raw materials and a corresponding growth in the art of coffee roasting, exotic coffees became accessible to a larger market. Although the prices are more than double those of grocery store brands, consumers are apparently happy to pay the premium for a perfect cup of coffee.

Some of the easiest profits are being made by coffee cart vendors! These independent entrepreneurs simply set up a push-cart and sell one or two blends by the cup to shoppers and strollers in malls and shopping centers. One of the cart vendors we spoke

with is making $350 per day working only four hours. Others claim to net $150 to $200 for five- to seven-hour days! With expenses running less than $25 per day, you can see the potential of operating several carts in good locations.

Opening a gourmet coffee and tea store in an accessible location will cost between $20,000 and $50,000, depending on the area and the amount of inventory you plan to stock. However, the majority of owners we spoke with told us they had made back their investment—plus a profit—within nine months to one year. By branching out into sales of coffee and tea accessories and related gift items, owners claim to be realizing gross income of $75,000 to $150,000 going into their second year.

This is a pleasant business that offers low-stress involvement with good profit potential. If you enjoy talking with people, you will find that a coffee and tea store is a natural conversation starter...people love to talk about a cup of coffee they had when they were visiting some strange and wonderful port of shore! Your knowledge about the blends you carry will give you an authoritative edge that will charm many customers. In addition, many owners told us how much they enjoyed watching the smiles on customers' faces when they tried a delicious new blend of coffee or tea.

Like any business, however, a gourmet coffee and tea store requires attention to the details of day-to-day operation, record-keeping, inventory control, and promotion, among other areas. By reading this business guide and following the suggestions outlined, it is conceivable that you could have your coffee and tea store up and running within a few months' time. Good luck!

1

THE GROWTH OF
SMALL BUSINESS

The 1980s were referred to as America's "entrepreneurial era." In 1986, more than 750,000 new businesses were created in the United States. In 1989, more than a million new ventures were started nationwide. The Small Business Administration (SBA) estimates that there are over 10 million home-based businesses in the United States today and, of these, more than 30 percent are owned and operated by women.

The recession of the early 1990s compelled suddenly unemployed people to start their own businesses in order to survive. An upsurge in part-time businesses has also fueled entrepreneurism in the nineties, with people in search of extra income working on profitable ventures during their off-hours.

More and more people are opting to leave their 9-to-5 jobs and stop "making someone else rich" to focus their energies on building a successful business of their own. For these people, it is no longer enough to spend nearly three-quarters of their lives working in a dead-end job, putting up with petty office politics, unappreciative employers, and long commutes just to bring home headaches and a small paycheck.

Today, the number of individuals who are self-employed is at its highest level ever, and based on your interest in this book, it's quite possible that you'll be joining the ranks of small business owners in the near future. It may be simply a dream right now, but that's how these businesses start.

You're never too young, too old, too busy,
or too poor to start a business.

Starting and operating your own small business is one of the most exciting and satisfying challenges you can undertake. There are no limitations on income potential when you're investing time and energy in your own enterprise. With the practical information provided in this book and dedication to your business goals, your chances for success are excellent.

What Is a Small Business?

The majority of businesses in the United States today are classified as small businesses. The definitions of what constitutes a small business run the gamut from the size of the overall staff (typically under 100) to the amount of assets or sales volume. However, in this book, small business is defined as one that is independently owned and operated.

The major benefit of this type of business is that you have the ability to make decisions quickly and act on them immediately. What typically bogs down big business is the number of people involved in the decision-making process.

Other advantages include the fact that small businesses can provide personalized service to the community or the market they're serving and the owner has the freedom, independence, and control to operate exactly as he or she chooses.

It's important to remember, however, that most major corporations, from Ford Motor Company and McDonald's to Mary Kay Cosmetics, started out as small businesses—as dreams.

It was because of basic business sense and a willingness to learn and adapt as their companies grew that Henry Ford, Ray Kroc, and Mary Kay, and thousands like them, steered their dreams into monumental financial successes.

Whether your goal is to supplement an existing income and to operate a solely owned home-based business from your garage, kitchen table, or spare bedroom, or to start a business that involves raising substantial capital, finding and setting up a commercial location, and hiring employees, you have the potential to enjoy an independent lifestyle that carries with it a number of rewards. And the rewards are as varied as the people who pursue them.

Entrepreneur
A person who undertakes an independent enterprise; one who has made the decision to go into business for him/herself.

Accepting the Challenge

Remember, these rewards do not come without hard work and the willingness to research and understand all facets of running your business. Many new businesses fail within the first few years.

Reasons given for the early demise of a small business range from lack of organization or management experience to undercapitalization, misunderstanding of the importance of advertising, inexperience in pricing products and services, lack of an overall business plan, improper hiring practices, and failure to meet the competition.

It isn't that someone purposely starts a business without having explored these areas. However, many times a person feels that his or her demonstrated expertise while in the employ of another can easily be transferred into a personal business. This is only partially true.

While it is imperative that you have these particular skills or talents—because selling them is what your ultimate success will be based on—it's equally important to understand how to sell them,

to know exactly what your profit margin is, and what steps you need to take to ensure the continued growth of your business.

Sounds easy, right? It really can be. But like anything else worth doing, starting your own business means careful planning. For example, you wouldn't consider taking a month-long vacation without doing some serious planning to ensure that the house was taken care of while you were gone, that you had made reservations for lodging, tours, and flights, and that you had converted your cash into traveler's checks.

There are so many aspects involved in running a business, it is vital to be prepared for any eventuality. Being prepared means being informed, so that when situations do arise, you know how to deal with them.

Is "Failure" Really Failure?

We have all heard stories about people who started their businesses on a shoestring and who became successful because of their sheer determination to make it work. It does happen, but these people are the exception rather than the rule, and in most cases have had experts standing behind them to give them guidance when problems came up. Others fail and, unfortunately, this often holds potential new business owners back. We hear and read about amazingly high figures related to so-called business failures.

According to a research project conducted by Albert Shapero, professor of the American Free Enterprise System at Ohio State University for many years, no one really knows the true failure rate of new business. The main reason for this is because there is not really a standard definition of failure in this case. He points out that a number of businesses close for a variety of reasons, many of which are not documented.

For example, in some cases the owners reach retirement age and have no one to pass the business along to; others shut down because the owners simply get bored; while still other entrepreneurs file a Chapter 11 bankruptcy, which basically gives them the opportunity to stay in business and continue operating under a court-approved plan, even though they become a statistic on the failure list.

The other extremely important aspect to consider when thinking about the benefits and risks of starting your own business is that having a business fail has never been a deterrent for true entrepreneurs. Many well-known business moguls failed at least once, and often more than once, before striking it rich.

Learning From Experience

In fact, almost anyone who has had a business fail will tell you that the experience was more valuable than anything they could have been taught in a business school and provided them with the knowledge they needed to start another venture successfully. This kind of determination is a valid qualification for self-employment and will pay off handsomely.

When you own your own business, you are responsible for everything. There will be times, such as when your accounts receivable are running sixty days late or the phone company puts the wrong number in your Yellow Pages listing, that returning to the 9-to-5 world will seem like a viable option.

This is where self-discipline, an unwavering belief in your product or service, and the determination to be your own boss will pull you through.

But, again, we can't stress enough the importance of planning, understanding basic business practices, being aware of consumer trends, and taking the time to develop, implement, and update goals to ensure success for your efforts.

What This Book Offers

This book provides you with the information you will need to start your coffee and tea store, offers techniques to help you with day-to-day operations, and provides actual case histories of people just like you who had a dream and, through planning and determination, were able to turn that dream into a successful reality.

In addition to focusing on aspects of the coffee and tea store, we cover such important business matters as

♦ Recognizing the entrepreneurial profile

♦ Taking our exclusive Entrepreneurial Quiz

♦ Finding the right audience for your business through easy marketing techniques

♦ Organizing for efficiency

♦ Recognizing legalities

♦ Addressing financial concerns

♦ Getting your home office up and running

♦ Charting your enterprise's growth

You will find specific how-to information on

♦ Advertising and promoting your business

♦ Finding capital

♦ Saving money on operating expenses

♦ Developing a simple bookkeeping system that will show you whether you're facing a financial crisis or realizing a profit

You're never too young, too old, too busy, or too poor to start a business. Owning your own business means taking advantage of our marvelous system of free enterprise. Earning a substantial living and, even better, realizing a profit for doing something you enjoy is the American dream come true.

The road to financial independence through self-employment is challenging and rewarding.

The opportunities for entrepreneurs have never been better. Armed with a solid product or service to sell, the determination to succeed and, most important, business know-how, there is nothing that can stand in your way.

Notes

Key Points

Personal Thoughts

Additional Research

2

THE ENTREPRENEURIAL PROFILE

Starting a business is one thing, making it work is another. We know that success in self-employment is largely the result of careful planning and an understanding of basic business techniques and formulas.

Start a business based on your expertise in a specific field and focus your involvement in an area that you thoroughly enjoy.

As many successful entrepreneurs claim, making money doing something you love is the best way to ensure a profitable future. It is always easier to address the inevitable business challenges that crop up when you are creating a product or providing a service that gives you a sense of pleasure and personal satisfaction.

Personality is a significant factor in determining what kind of business to get involved in, the way you will eventually set up the legal structure (sole proprietorship, partnership, etc.), and how you will run the business on a day-to-day basis. For example, if you are planning to start a business that is based on your artistic

or creative abilities, it is possible that your personality is not suited to the very important aspect of sales. But without strong selling abilities, there is a likelihood that your goal of distributing, for example, your hand-carved wooden boxes nationally will not come to fruition.

This isn't to say that you should decide against going into business for yourself. It simply indicates it would be in your best interest to join forces with someone who does have strong selling skills, who believes in the product as much as you do, and who will work with you toward a common goal.

On the other hand, if your personality is geared to working with people, it is a good idea to consider a business that will emphasize this ability, such as developing seminars or workshops based on your area of expertise, providing independent counseling or tutoring, or providing a service such as gift basket design or secretarial help, which depends on your interaction with people on a one-to-one or a group basis for success.

Research shows that in order to address the many and varied situations that arise in business ownership, the true entrepreneur should possess the following personality traits.

Ten Important Entrepreneurial Traits

1. Self-awareness
2. Self-motivation
3. Courage
4. Confidence
5. Patience
6. Decisiveness
7. Experience
8. Knowledge
9. Perseverance
10. Drive

1. ***Self-awareness.*** Know who you are, what you're good at, and what you like to do. Start a business that is based on your expertise in a specific field and focus your involvement in an area that you thoroughly enjoy.

2. ***Self-motivation.*** Also known as drive, self-motivation is one of the most important personality traits of successful entrepreneurs. This is the characteristic that gets you

going and keeps you moving when you are in business for yourself. It's what helps you to keep turning out those craft items, upgrading your technical skills, or developing new and improved promotional techniques when business is slow. It's what gives you the tenacity and confidence to call on a potential client even though you've heard "No" three times.

Self-motivation is also what helps you overcome the fears and concerns that inevitably arise when you own your own business. It is the main ingredient that has spurred on people who have achieved success despite drawbacks, such as minimal capital, lack of education, or limited experience.

People with a high degree of self-motivation see the greatest obstacles, such as learning a new aspect of business management, as a new and exciting challenge to overcome. If you've ever undertaken a project without fully understanding the mechanics involved in performing the task or knowing what the outcome would be, you were operating on self-motivation—the conviction that you would be able to learn whatever was needed to accomplish your goal.

And regardless of the outcome of the project, you undoubtedly gained more experience and knowledge than you had before, which only works to increase your sense of motivation to handle new challenges.

3. *Courage.* Courage or the willingness to take risks is a valuable trait when striving for success. A successful person might say something like: "I don't know how I did it; I just made a phone call and asked for the money I needed." It was more than luck that made it possible for this person to raise the needed capital to get a business off the ground; it was the willingness to take a chance and hope for a positive response to the request.

The owner of a small cabinet-refinishing business said, "I always figure that the worst thing that can happen is someone will say no, so it never hurts to try." In the game of business, you must be willing to take chances. Even if

you don't get exactly what you want every time, the odds are good that if you feel strongly about what you need, you will get it. But you have to ask!

4. **Confidence.** The age-old philosophy of positive thinking is a major step in the direction of success. By behaving as if you already are a success at what you do, it follows that you will be, and your customers will believe it, too. A confident attitude is one of the most appealing traits you can exhibit to prospective clients, for it lets them know that they will be getting the best their money can buy.

You have to accept whatever comes, and the important thing is that you meet it with courage and with the best you have to give.

Eleanor Roosevelt

5. **Patience.** When you own your own business, there will be times when you feel like the roof is caving in, especially when your suppliers seem to be taking their own sweet time in fulfilling an important order or when a customer's demands seem to be unrealistic. Although you may be able to hurry the supplier along, you must remember that your customers are always right, since they are the ones who will make or break your business.

If you are aware that patience is not your strong suit, develop a stop-gap exercise to use at times when coping is a definite necessity. Whether you count to ten before saying a word, visualize a pleasant scene, or repeat a secret phrase to yourself when tension is running high, it will be to your advantage.

6. **Decisiveness.** Business has been described as a process of making one fast decision after another. Often, a decision has to be made immediately, on the spur of the

moment. In those instances, you should go with your intuition and trust that you are doing the right thing.

However, if you are the type of person who prefers to analyze your options, weigh all the factors, and make decisions slowly, then that is what you must do. It will keep your confidence intact and ensure that you're taking the right action. Again, careful planning will help you predict many of the decision-making situations that arise in your business. As time goes by and you grow more comfortable in your role as business owner, you will undoubtedly find yourself making quicker decisions.

7. **Experience.** A primary reason some businesses fail within a few years of start-up is lack of business experience. Whether or not your experience is directly related to the business you are planning to start, it is a key component for growth.

If you feel you do not have enough business experience, there are several avenues you can take before starting your own enterprise. Returning to school for specialized courses is one consideration. Most community colleges and adult education facilities offer classes and seminars in business start-up and maintenance, and there are hundreds of home-study or correspondence courses that will, in many cases, give you official certification in your field.

However, your best solution is to take a job in the field you're interested in. By asking questions about all the aspects of the business, you will gain experience, get paid for learning, and find out whether this is really what you want to do before sinking money, time, and energy into the enterprise.

8. **Knowledge.** Experience in your field is invaluable, but it is not enough. You have a better chance of succeeding by knowing as much as possible about such vital factors as business basics, the product or service that you want to sell, the industry you are entering, your competition, and the local, regional, state, and/or national markets. Do your homework.

9. **_Perseverance._** One of the adages you will hear time and time again when talking to entrepreneurs is that perseverance is 90 percent of the battle to succeed.

 If you are like the majority of new small business owners, the entire staff and support system for your venture is probably _you._ Making a dream come true can be a lonely task, especially when you are just getting started, and ensuring that it works often means little rest or relaxation. You must be willing to persevere during the rough times, to hang in there during the slow periods, and to maintain your belief in your product or service—even when it seems like no one else in the world knows you exist. It has been written that through perseverance the snail reached the ark. So it is with success!

10. **_Drive._** You must believe in your goal and be determined to succeed. The successful businessperson does not view obstacles or stumbling blocks as problems, but rather as challenges. To make your business a success, you need to be totally committed and willing to work long hours, especially in the beginning. If you do not have the time or commitment to dedicate to your dream, then you are better off working for someone else.

The Entrepreneurial _A_ to _Z_ Appraisal

Owning a business calls for the ability to handle different situations with confidence. The following self-appraisal quiz has no right or wrong answers. It is designed to help you determine certain personality traits, attitudes, and qualifications that will benefit you in your venture. Upon completion, use the appraisal as a starting point for discussions with friends and family members about your business profile. Acknowledging your strong and weak points will help you prepare for your role as an entrepreneur.

Entrepreneurial Quiz

Use the letter *S* for strong or *N* for needs improvement beside the characteristics listed below. Give yourself sufficient time to analyze each trait.

Achievement: I have a strong desire to be successful in my chosen business venture.

Belief: I have faith in myself and the service or product I am specializing in to build my business.

Creativity: I am able to address situations in imaginative and innovative ways to reach my goals.

Discipline: I am self-motivated and can handle necessary tasks, whether or not I enjoy them.

Efficient: I am organized and can arrange my priorities or change my work methods as needed for maximum production.

Friendly: I am genuinely interested in people and enjoy my interactions with them on a day-to-day basis.

Goal Oriented: I have a tendency to set my sights on preset goals and will work hard toward achieving them.

Health Conscious: I am aware of my physical abilities and have the insight to work smart in order to preserve my health.

Independent: I am able to work alone, if necessary, and am willing to be responsible for my own actions.

Judgment: My conclusions about people or situations are generally accurate.

Knowledge: I have solid experience in my field and have spent enough time in a professional business setting to know the ropes.

Leadership: I am able to direct people effectively while instilling confidence and loyalty. _____

Maturity: I am willing to work toward long-term goals and do not get upset by the inevitable minor setbacks. _____

Networking: I am willing to develop associations with other entrepreneurs for professional support in my venture. _____

Optimism: I am able to see what is right about a situation and to explore its potential to the fullest. _____

Positive Attitude: I am convinced that I can accomplish anything I set my mind to, and rarely entertain negative thoughts. _____

Questioning: I am not afraid to ask questions to get the information I need to expand my knowledge. _____

Resourceful: I am able to find ways to accomplish just about any task I must do. _____

Sales Ability: I am able to present information about myself and/or my business in a convincing yet honest manner. _____

Tolerance: I am able to handle stressful situations with a positive and realistic attitude. _____

Undaunted Spirit: I am unafraid of the unknown. In fact, I enjoy a challenge and accept the consequences of my actions. _____

Venturesome: I am not afraid of hard work to reach my goals and enjoy finding new, positive ways to handle troublesome situations. _____

Well Balanced: I generally maintain a sense of humor when things don't work out as expected. _____

eXpressive: I am able to express ideas and feelings clearly and logically, both orally and in written form. _____

Youthful Nature: I am capable of tackling work with enthusiasm and a high level of energy.

Zest: I look forward to enjoying my business, the people I will be dealing with, and the resulting fruits of my labor.

Scoring

Although this is merely a tool to provide you with information about your entrepreneurial profile, there are immediate clues to your future as a business owner in the responses you have given.

Add up your number of *S* responses and read the following interpretation that corresponds to the number you scored.

♦ 15+: There is a good possibility that you have been involved in your own business in the past or, at least, have worked in a managerial capacity for someone else. You have the positive personality traits required to be a successful business owner.

♦ 8–14: You are basically a positive and directed person and should not have any problem with improving certain areas to increase your personal business success potential.

♦ 0–7: This is an indication that finding a complementary business partner who can support your goals may be an option worth considering.

Notes

Key Points

Personal Thoughts

Additional Research

3

CHOOSING THE RIGHT BUSINESS

Selecting the type of business to start is one of your primary steps in becoming an entrepreneur. It is not predicated on luck, although we've all heard about someone who just fell into a certain business opportunity and was able to make it a success. For the most part, however, it is a conscious decision based on personality, abilities, and interests.

You want a business that suits your goals and personal lifestyle so it will be enjoyable. You must remember that you will be spending a lot of time and energy getting it off the ground, so it is important that you are doing something that you like.

Business Capabilities Application

If you have not yet decided on the type of business you want to own and operate, it is advisable to prepare a *Business Capabilities Application* (BCA). Similar in nature to a job application, your personal BCA will give you a good idea of skills you have gained through education, jobs, and hobbies. Thus, it can be of great importance in helping you find the right opportunity.

*Experience is not what happens to a person. It is
what a person does with what has happened to
them in the past.*

<div align="right">Anonymous</div>

Your Business Capabilities Application is a personal research tool,
so it should be written out after serious consideration and analysis
of the categories indicated below. Give yourself plenty of time to
work on the BCA. Once you have written your answers, put them
aside for a few days and then review them again to see if there is
anything crucial you might have left out the first time.

Education

Be sure to include specific courses that have had an impact on
career decisions, as well as any courses that might prove useful in
the running of your business (art, bookkeeping, management
training, etc.) Also, indicate seminars or workshops you have
attended that increased your knowledge of specialized areas of
business.

Work Experience

Outline the jobs you have held, what your exact responsibilities
were, and what you enjoyed most about each job. List any work
habits and character traits that proved to be helpful in accomplish-
ing your goals while holding these positions.

Health Background

This is basically necessary only if you have a history of illness or
any physical limitation that could affect your choice of a business.
For example, if you have chronic back pain, you would not want
to own a business that requires heavy lifting or being hunched
over a worktable for extended periods of time.

Hobbies

List hobbies you are currently pursuing and those you enjoyed in the past. Turning a hobby into a business is one of the most popular and successful ways of capitalizing on skills.

Volunteer Experience

Include everything from political campaign support to chairing the PTA book drive. Volunteers typically delve into many areas that provide solid experience for running a business, such as telemarketing, working on a budget committee, organizing events, or serving as a spokesperson. Outline your volunteer experiences carefully, indicating details of involvement.

"If there is a faith that can move mountains, it is faith in your own power." Austrian author Marie Ebner Von Eschenbach wrote this phrase back in the mid-nineteenth century, and her words are as true today as they were when she first suggested the power of positive thinking.

Other Interests

Use this category to make note of any additional skills and interests not included in the other categories. Those might include such areas as entertaining, a special ability with animals, children or seniors, your touch in the garden, a flair for flower arranging, etc.

Personal Goals

Indicate the five most important personal goals that you would like to accomplish over the next two years. These might include increased self-esteem, learning to complete projects, leaving a boring job, finding a new career, or having the opportunity to express a creative ability. Personal goals also include factors that cross over into business goals, such as earning enough money through self-employment to purchase a new home within five years or to gain experience in the business that enables you to write a book or develop a seminar program.

Business Goals

Thinking in terms of owning your own business, write down at least five goals you would want to achieve, such as independence, financial security for retirement, national recognition, or recognizing a need and filling it.

Writing your skills, experiences, interests, and goals down on paper may provide just the spark needed to help you zero in on the perfect business. Once you have completed the Business Capabilities Application and reviewed it several days later, go back through and review the information with a colored marker in hand. Underline or circle any details that strike you as being something you particularly enjoyed doing.

You should see a pattern starting to emerge, such as an interest in working with people, or perhaps a preference for working outdoors. You are now halfway home to finding a business suited to your personality and lifestyle. It has been proven that people who *enjoy* what they are doing are more successful than those who participate in a business simply because it seems like a good way to make money.

Looking at the Possibilities

The next step is to make a list of all the possible business opportunities that match the information on your Business Capabilities Application. Consider how much time and energy you are willing to put into the business and what your expectations are for the short term, such as securing your position in the local business community, as well as for the long term, i.e., a flexible schedule based on being in the economic position to hire employees.

Say, for example, you like to sleep until noon, enjoy preparing exciting new recipes for friends, and have minimal start-up capital, but you expect to see financial rewards from your business. Your first idea for a business might be to open a restaurant.

However, owning a restaurant generally means getting started at 6 A.M. to prepare for lunch crowds. In addition, the expense of setting up an average family-style restaurant is conservatively estimated to be $50,000...and that is before you start hiring serving personnel and launching a major advertising cam-

paign to let the local population know that you are now in business.

Taking these factors into account, it would appear that, because of your sleeping habits and low cash resources, restaurant ownership would be out of the question. However, other possibilities to consider would be operating a catering company from your home, developing a specialized food product to manufacture and distribute, or creating a unique meal delivery service in which you prepare the meals and deliver them to working families, home-bound seniors, and other clientele you obtain through a carefully designed marketing program.

Narrowing the Choices

Once you have developed your list of possible businesses, the next step is to rate them in order of priority, according to your personal and business goals. Make a list of the possibilities running down the left side of a sheet of paper and note your goals across the top. Give each business a rating of one to ten as it relates to the goals, then add up the scores. The businesses with the highest scores are the ones you should explore further as having solid potential.

Check your listing of goals once more to insure that you have included everything. If so, you can now start to research the strong contenders in your search for the business choice that best matches your needs and desires.

When conducting your research, ask friends and family members to provide their thoughts on which of the businesses they can see you operating successfully and talk to the owners of similar businesses and ask them to share their experiences. Ask for input from experts, such as your attorney, representatives of SCORE (the Service Corps of Retired Executives) through the nearest Small Business Administration office listed under U.S. Government in the front of your telephone book.

Finally, trust your intuition about the top contenders. Since you are the one who will be living with the business day in and day out, your final selection must be one that satisfies your goals, meets your criteria for growth potential, and features the kind of management and production tasks that you can easily handle.

Notes

Key Points

Personal Thoughts

Additional Research

4

STARTING A COFFEE & TEA STORE

If you are like many people who enjoy coffee and tea, you think of these beverages in relation to taking a break–relaxing with the newspaper in the morning, stopping for a breather in the middle of the day, or finishing off a perfect meal. Although Europeans have refined the coffee or tea break into an art form, Americans' interest in specialty coffees, teas, and associated drinks is a relatively new phenomenon.

According to The Specialty Coffee Association of America, the coffee industry grew from a fledgling industry started in the early 1960s to one that had achieved sales of $45 million by that decade's end to one that is expected to approach $5 billion by the turn of the century. Of this projected figure, there is an estimated $3 billion in sales for home consumption and $1.5 billion estimated for retail food service sales of specialty coffee beverages.

The Specialty Coffee Association of America sees the following trends:

1. Niche product categories will continue to grow and strengthen.

 ♦ Decaffeinated coffees will be available in a broader range as decaffeination methods improve quality.

- ♦ Flavored coffees, such as top-selling vanilla- and nut-flavored combinations, will increase in market share.
- ♦ Organically grown coffee sales will rise, as available, due to consumer health awareness.
- ♦ The specific geographic location, not just the country of origin, will increase as a marketing angle.
- ♦ The popularity of dark roasts for espresso-based beverages will continue.

2. Coffeehouses, espresso bars, and espresso carts will be the distribution channel with the fastest growth rate.

3. There will be a rise in "micro-roasteries" comparable to beer microbreweries.

Overview of Coffee & Tea Stores

This specialty retail operation capitalizes on the nation's love of coffee and tea. It has excellent profit potential, and growth is predicted to continue at a steady rate. It is a perfect owner/operator business.

"Shoestring" Start-up Investment:	$2,000
Average Start-up Investment:	$15,000
	to $20,000
High Start-up Investment:	$40,000
Break-even Point:	One year
Average Annual Gross Revenues:	$25,000
	to $60,000
Potential Annual Gross Revenues:	$100,000+

Truly dedicated coffee lovers prefer to buy whole beans, which means they will need a coffee grinder. Tea connoisseurs will not settle for the prepackaged filter bag variety of tea. They prefer loose-leaf tea to ensure full body, and tea cozies and strainers are a necessary part of the preparation process.

Beyond the Realm of Relaxation

Coffees and teas have become more than beverages that offer irresistible aromas and delectable flavors; they are rapidly being recognized as the perfect all-occasion gift. This is a major profit center for entrepreneurs. While most people would not dream of purchasing a can of coffee at the supermarket and giving it to someone as a gift, many would certainly go to a specialty coffee store, purchase a pound of Vanilla Viennese in an attractive paper bag, put a colorful ribbon on it, attach a greeting card, and present it as they would a bouquet of fresh flowers or a bottle of rare wine.

Profit-Makers to Sell in Your Coffee/Tea Store

- ◆ Baked goods and desserts
- ◆ Beans
- ◆ Books on the history of tea and coffee
- ◆ Coffee grinders
- ◆ Coffee makers
- ◆ Collectible china tea cups
- ◆ Cookies
- ◆ Demitasse cups
- ◆ Espresso brewers
- ◆ Flavored syrups
- ◆ Loose-leaf tea
- ◆ Mugs, decorative or unusual
- ◆ Prepacked gift assortments of various blends
- ◆ Sandwiches
- ◆ Spices and flavored sugars
- ◆ Spoons (demitasse, iced tea, etc.)
- ◆ Teapots
- ◆ Tea cozies, strainers

Although many people do not take the time to learn the nuances of coffee types, brews, and roasts, they know that differences exist. They are aware that coffee purchased in a specialty store always smells and tastes better than the prepackaged coffee that they bring home from the supermarket.

You can capitalize on the growing awareness of and appreciation for quality coffee and tea in several ways. It is possible to make a very comfortable living selling aromatic cups of freshly brewed coffee from a pushcart or to realize excellent profits by setting up a full-scale coffee and tea store in a commercial location. You may decide to sell coffees and teas only or to include complementary items in your stock as well—accessories, spices, gourmet cookies, and chocolates.

A shop situated in the appropriate location and modeled after an outdoor cafe invites potential customers to have a cup of cappuccino, espresso, or private blend and to sit and watch passersby. The aroma of freshly brewed coffee is the sensory magnet that pulls coffee lovers into the shop and fills them with excitement about having to choose between French Vanilla or Mocha Java.

Don't neglect to create an interesting atmosphere in the store's interior. One advantage of this business is that you can display a great deal of your inventory to create an appealing environment that encourages in-store as well as take-home purchases. Load shelves with coffee accessories such as attractive coffee mugs and canisters and fancy cappuccino, coffee, and espresso makers. Fill glass cases with different coffee beans and label with the country of origin. The noises and smells of the business are also appealing—coffee grinders are whirring, the espresso machine is steaming and hissing, and the wonderful aromas fill the air. Coffee and tea lovers are definitely people whose senses of smell and taste are keen and in need of satisfaction. Your store needs to offer items that smell, taste, and look good!

You want to appeal to as wide an audience as possible, by maintaining a varied inventory. At the same time, you want to be selective in the choices you stock in the store to avoid creating customer confusion as well as to keep overhead expenses within a reasonable range. The fewer suppliers to deal with, the easier your ordering will be.

Unlike supermarkets, where the markup is small and a net of 2 percent on gross sales is considered good profit, a gourmet coffee and tea store can mark items up from 30 percent to 100 percent or more. Selling coffee by the cup has a high profit margin. And consumers are willing to pay a premium for many espresso-based beverages that are difficult to prepare correctly at home. Coffee is a commodity with varying prices, thus, you will need to vary your prices as well. Overall, though, you can look forward to net profits of 15 to 20 percent.

There are few hours in life more agreeable than the hour dedicated to the ceremony known as afternoon tea.

Henry James, *Portrait of a Lady*

Whatever path you choose, it is estimated that gross annual income in this growing business can hit $40,000 to $75,000 by your second year. Adding mail order could bring in an additional estimated gross annual income of $25,000 or more depending on your ability to hit the right market.

After the coffee, things ain't so bad.

Anonymous

Qualifications

What do you need to start a coffee and tea store, to operate a beverage pushcart, or to get into the mail-order end of the business? Obviously, it helps if you love coffee and tea! And people! A thirst for knowledge as well as for a well-roasted brew is going to enhance your ability to do well. The main requirements for success are a desire to learn all that you can about your products and, of course, to be able to address the day-to-day tasks involved in running a business.

A palate for distinct flavors is helpful, but don't worry if you can't really tell the difference between a Colombian and a Costa Rican blend at this point; your skills will develop. Many people can't tell the difference among types of coffees simply because they have not paid attention to them. Once you read about roasts, grinds, and brews and begin to take notice of them, their subtleties will become meaningful to you.

You also will want a basic understanding of marketing techniques. Visiting coffee and tea shops, or even other types of gift shops will be helpful, and taking a business course or two couldn't hurt if you've never run your own business before. You might try working in a coffee and tea store, if there is one in your area, to learn the business. There is nothing like firsthand experience.

Profit Potential

The profit potential in this business is limited only by your ability and desire to create different ways to sell coffee, tea, and other store merchandise. According to several store owners, the greatest profits are made when merchandise is sold as gifts. One reason is that customers will typically spend more money on a gift for a friend or loved one than on a purchase for themselves.

We would love to tell you that you can start a specialty coffee and tea store for $100, but it just isn't so. The store proprietors we spoke to all suggested minimum opening capital for a 500- to 1,000-square-foot store in a good location to be somewhere between $30,000 and $40,000. A few said it is possible to start for under $20,000 if you rent a small space in a well-trafficked shopping area.

Simple and Small

It is best to start small with a low inventory so that your financial risk is limited. The general public thinks of entrepreneurs as extraordinary risk takers. The truth is that successful entrepreneurs are typically creative but careful.

The easiest way to begin selling coffee is from a pushcart in a busy mall. In this situation, all you need is the following:

1. Topnotch espresso machine
2. Espresso coffee blend
3. Milk, cream, sugar, honey
4. Coffee stirrers, cups, napkins
5. A distinctive outfit, such as a tuxedo
6. Money apron

One successful "coffeepreneur" we interviewed got a license to set up his cart in a busy shopping mall that attracts thousands of people every day. Many of his customers are employees who flow out of dozens of nearby office buildings during lunch hour.

He purchased a secondhand copper espresso machine for $4,000, paying half down, with the balance due in two months. He also bought a small but sturdy used pushcart from a former flower vendor for $300 and spent $25 on paint to redecorate it. Sugar, honey, cups, napkins, and other miscellaneous items came to $50. Off he went to the shopping mall, stopping every day in the supermarket for fresh cream and milk.

He sells cookies he bakes himself, for which he charges $1 a piece. A cup of espresso goes for $1.50. His busiest times are from 10:30 to 11:00 A.M., when people from the surrounding shops run out for a fast cappuccino, and from 12:00 noon to 3:00 P.M., when people grab fast lunches and afternoon snacks. On Saturdays the bulk of his business comes from mall shoppers. This coffee vendor grosses between $125 and $250 per day, working five to six days per week.

The Mail-Order Business

More coffee is sold from commercial locations versus through mail order, but some companies pursue the latter and have made good money at it. Their claim to success is knowing how to reach those people who love gourmet coffees and teas but who live in areas where these products are hard to come by.

Write to one or more of these mail-order companies to see how they set up their order forms. Call or write and order a pound of coffee to check out the company's system, how long delivery takes, packaging, and other specifics of mail order that you should examine before getting into it yourself.

Notes

Key Points

Personal Thoughts

Additional Research

5

START-UP BASICS

Having decided that you are ready and able to accept the challenge of starting your own business, it is necessary to take a look at your overall financial picture. Even if you have a healthy savings account, or feel that you can start your business with a minimal capital investment, examining your personal financial situation will help you identify ongoing expenses.

The easiest way to estimate exactly how much money you will need to start your business and to cover expenses (including personal living expenses for the first six months) is to prepare a *Cash Flow Statement* and a *Projected Expense Statement.* Samples are provided in this chapter.

Preparing the Projected Expense Statement will give you a fairly accurate picture of what it will cost to open the doors and will indicate how much income you must generate to realize a profit. Another advantage of creating these charts early in the game is that two of the required documents will already be prepared when you want to explore funding options, then you will only need to update them.

The first step in examining your financial situation is to ask yourself the following questions:

1. Do I generally pay my bills on time or do I wait until my creditors start sending me collection notices?

2. Have I regularly reconciled my bank statement so I know how much money I have in my checking account at any given time?

3. Is my philosophy "If I've got it, I spend it" or do I carefully plan how I am going to use my income?

4. Have I ever developed a personal budget so I know how much money is coming in, how much is going out, and what I have left over?

These are important aspects of your financial personality that will be helpful to understand when running your business. As your business and subsequent involvement with financial matters grow, it will be vital that you have a handle on your philosophy about money. And there is no time like in the beginning, when your business concept is being formed, to start learning.

The Cash Flow Statement

Use the Cash Flow Statement at the end of this chapter to determine your personal living expenses for the past three to six months to help you gauge what you will need to survive during the early stages of your business.

The easiest way to complete the statement is to use your checkbook register or sales receipts as research tools. If your expenses are relatively consistent from month to month, you should be able to get an overview by analyzing one month. A more accurate picture will emerge if you break down income and expenses for three to six months to account for periodic payments, such as taxes, insurance, and seasonal spending.

Using the samples provided, fill in the amounts in each category from your checkbook register or receipts. Use a separate sheet for every month that you are analyzing. For miscellaneous spending, a standard calculation is 5 percent of monthly income.

Add up each month's expenses, total them all, and then divide that figure by the number of months you are analyzing. This will give you an average monthly expense figure.

Follow the same procedure for income. You can then subtract your expenses from your income to see where you stand. If you have computed your figures accurately, you might run across a few surprises. It isn't unusual to discover that we spend more money than we realize, often on unnecessary purchases. You may be able to see some areas where you can cut back.

The main points, however, are that you now know

1. How much or whether you can afford to invest your own money in your new business

2. What it costs for you to live comfortably, which will help you set income goals for the business

A six-month cost projection will be a good indicator not only of expenses near-term, but the kinds of profits you can expect as the business gets underway.

Start-up Costs

Every business owner has specific standards about how he or she wants to run the operation. One person may feel perfectly comfortable waiting to order business cards until there is a profit. Another wouldn't dream of opening the doors without cards, brochures, and letterhead already printed.

You will have your own ideas about what you need before opening your business. Then you must find out what it will cost and, if at all possible, prepare the *Start-up Costs Worksheet* at the end of this chapter.

It is also advisable to figure how much it will cost to run the business for three to six months, using the sample Projected Expense Statement. A six-month projection should give you the opportunity to determine your profits down the line.

Preparing the Start-up Costs Worksheet and Projected Expense Statement involves conducting some research. For example, to estimate the cost of business cards or letterhead stationery, contact several printers or copy shops in your area and obtain quotes. Call the local newspaper for prices on different types of ads, including display and classified.

An insurance agent will be able to give you an estimate on liability coverage. Check with the telephone company for information and rates on installing a phone line.

You can also start to shop around to find the best prices on office supplies, equipment, and materials you will need to conduct business.

After you have completed your research, incorporate the information on the blank charts. Obviously, some of your figures, such as those for telephone expenses, will be guesstimated. However, the final figure will give you a good idea of start-up and operational costs for the first six months.

Utilizing the same theory, you can develop a *Projected Income Statement* (worksheet not provided), drawing from industry figures available through your trade association or other research sources. This would include all income realized from cash sales, collection on outstanding invoices, credit card sales, and miscellaneous income. By subtracting your total expenses from total income, you will get a clear picture of projected profit or loss.

All of these statements will be requested by loan officers, venture capitalists, and the Small Business Administration if and when you apply for a loan. This kind of paperwork is required to ensure that you have basic business knowledge, a commendable track record, and are serious about your venture. You will also be required to fill out a personal financial statement, available through the lending institution, especially if you are the sole owner or a general partner in the business.

General Office Equipment

There is no reason to pay high prices for office equipment and supplies. Everything needed to outfit your office can be purchased for at least 50 percent below standard retail prices either on sale

or through wholesale clubs, office supply discount stores, or wholesale catalogs.

Adding machine. Many computers include a built-in calculator and keypad. The advantage of an adding machine is that it provides a paper record of your figures, while a calculator provides only a temporary display. This will be especially important when you are doing your daily cash totals.

Bookcase and inventory shelving system. Your printed matter (brochures, takeout bags and boxes, business cards, etc.) will take up quite a bit of space, especially since you will want to order enough to last a while. There might also be sales receipt tape or pads, purchase orders, company stationery, promotional material, etc. A bookcase or shelving system will accommodate these items and keep them organized.

Computer considerations. Initially, a computer is certainly not a necessity, but it makes inventory control ten times more efficient. As your business grows, the administrative duties will increase and require the preparation of inventory charts, customer databases, and other information.

When your budget warrants it, start investigating prices on computer equipment and supplies that you will eventually need to handle your administrative responsibilities efficiently.

As your business grows, you will find computerization to be the answer to time and workload constraints. You can find more details in Chapter 6, "Computerizing Your Office."

Desk and chair. You're going to spend a significant amount of time either sitting in front of your computer or sitting at a desk doing paperwork and making telephone calls. Therefore, it's extremely important that you invest in the right desk and chair.

Not only will the wrong furniture give you back and neck aches, studies have proven that it can cause extreme fatigue and a decrease in overall productivity.

Depending on your height, you will want your computer keyboard desktop to be about twenty-six inches from the ground; writing surfaces are generally thirty inches high, so you might want to look for an adjustable computer desk that accommodates both tasks comfortably.

Even more important is the chair. It should provide support for the base of the spine, a backrest no higher than your shoulder blades, and a seat that can be raised or lowered easily. A solid seat and rounded cushion to prevent the restriction of blood flow in the legs is recommended. The lumbar support and armrests should also be adjustable since everyone's body is proportioned differently. The best way to find the chair you need is to visit an office furniture store and try out several until you find one that feels comfortable.

Filing cabinet. An absolutely essential item is a filing cabinet to keep all of your papers in order. It is not enough to keep a ledger of income and expenses. You must also save all invoices, receipts, canceled checks, purchase orders, equipment warranties, copies of delivery slips, coffee blend recipes, etc. If you employ help, you must keep payroll records as well.

Of course, much of this information can be input and maintained directly on your computer, but there is still need for a simple filing system for "hard copy" and those paper odds and ends that must be saved for income tax purposes. All you need is a standard two- or four-drawer filing cabinet to begin with. Be sure, however, that you don't fall into the trap of buying your filing cabinet through a discount department store—typically these are substandard brands and frequently are not properly weighted. We've heard more than one story about drawers falling out or the whole cabinet falling over on some poor unsuspecting person.

Telephones. In a commercial location, getting a business telephone will be one of the first steps in setting up your business. Although it is still possible to rent telephones through certain telephone companies, the best bet is to buy your own. They are inexpensive, come in a variety of styles and colors, and are generally covered by a warranty.

Special Equipment Needs for Your Coffee and Tea Store

Any piece of equipment you use you will buy, lease, or rent. (The information on these options provided in Chapter 6, "Computerizing Your Office," applies equally well to most other major equip-

ment purchases besides computers.) The best advice is to purchase the equipment if you have the capital, since it is the most economical way in the long run and you can depreciate the equipment on your income tax; lease equipment if you do not have enough capital to purchase, or if you want to establish a credit rating. Before committing to a purchase or lease, you may want to rent equipment to try out while waiting for a loan to buy.

You will need various specialized pieces of equipment to brew, grind, and make various beverages. Since needs will vary depending upon the scope of your operation as well as your financial resources, this book will not attempt to recommend specifics for these items. Instead, we suggest that you first identify your needs, then contact various suppliers and obtain several price estimates before making a major purchase. The Appendix includes a listing of some suppliers of these items.

After your coffee grinders and brewing equipment have been installed, it must be inspected by fire underwriters who will determine its safety. Be sure that you check these appliances periodically and make sure that they are clean and the heat controls are working properly. Here is a list of things to check for.

1. Plugs fit properly (not loose or cracked).

2. Cords not frayed, twisted, or otherwise in poor condition.

3. If a piece of equipment is not working, did you check the fuse box?

4. An electrician can repair broken wires in electrical heating units. This would save buying a new unit.

5. When using water to clean around electrical equipment, make sure no water gets in the heating element. This can be extremely dangerous.

Cash register. You can buy a fancy computerized cash register that provides built-in inventory control or settle for a simple unit that allows you to keep track of income. The option you choose will be based on your available capital.

Display equipment. Whether you choose to display products on high-tech shelving or in antique armoires, you must figure on spending money on this vital necessity. The decorating scheme of

your shop is going to serve as an important promotional device, as well as inviting customers to browse and buy.

Enhanced packaging equipment. At some point, you will be asked to ship a gift somewhere out of town or even out of the state. Although you will most likely be using one of the major shipping companies, such as Federal Express or UPS, you will still be required to pack the purchases in boxes for delivery. There are some things like special tape dispensers and shrink-wrap machines which, while generally expensive, might pay for themselves when volume warrants their purchase.

Lighting. This is the most overlooked and yet most important equipment in retail stores. In your coffee and tea store, consider fluorescent overhead lighting augmented by freestanding lamps or track lighting for a soft, yet friendly atmosphere.

Refrigerator. Perishable products such as milk and cream will need to be kept under refrigeration. The size of your unit will depend upon what foodstuffs you plan to offer, but don't underestimate your storage needs.

Scale. You will be selling coffee by the pound, so you will need an accurate scale for weighing out customer purchases.

Signs. The type and the size of sign you put outside your store may be dictated by the lease. Be sure to check before investing from $200 to $500 in something you will not be permitted to use.

Supplies/Inventory

Basic Necessities

- ◆ Beverage stirrers
- ◆ Coffee, assorted whole beans
- ◆ Coffee bags
- ◆ Coffee pots
- ◆ Condiments, if food is served
- ◆ Cream
- ◆ Filters
- ◆ Freezer bags
- ◆ Honey
- ◆ Milk
- ◆ Mugs and cups
- ◆ Napkins
- ◆ Plates, if food served
- ◆ Salt and pepper, if appropriate
- ◆ Sugar
- ◆ Sugar substitute
- ◆ Tea, assorted loose-leaf
- ◆ Teapots
- ◆ Utensils

Optional Supplies

(Those that are a repeat of any items listed above are intended for sale.)

- ◆ Books and magazines on coffee and tea
- ◆ Candies
- ◆ Cappuccino makers
- ◆ Coffee grinders
- ◆ Coffee makers
- ◆ Coffee scoops
- ◆ Espresso makers
- ◆ Filters, paper and permanent
- ◆ Gift baskets
- ◆ Greeting cards
- ◆ Herbal teas
- ◆ Jars
- ◆ Mugs and cups
- ◆ Spices
- ◆ Tea cozies/strainers
- ◆ Teapots
- ◆ Tea towels
- ◆ Whistling tea kettles

Espresso Pushcart Start-up Costs

Espresso machine	$4,000.00
Pushcart	300.00
Business license	50.00
Espresso coffee	100.00
Milk, sugar, honey	20.00
Stirrers or spoons, plastic/paper cups	30.00
Distinctive serving outfit	50.00
TOTAL	$4,550.00

Specialty Coffee Store Start-up Costs

Espresso machine	$4,000.00
Coffee grinder	1,000.00
Sink	500.00
Drainage system	1,000.00
Counters	2,000.00
Health permit	500.00
Covered flooring/installation	2,000.00
Refrigerator	300.00
Storage space	500.00/month
Rent	1,000.00/month
Coffee, coffee makers, teas, teapots & other stock, plus serving supplies	10,000.00
Scale	500.00
Glassware (storage and display)	5,000.00
Track lighting	2,000.00
Electrician to hook up lighting	2,000.00
Carpenter to build counters	3,000.00
Business license	500.00
Miscellaneous	1,000.00
Cash register	200.00
Desk, filing cabinet, calculator	500.00
Utilities and telephone deposits	150.00
Insurance (3 months)	250.00
Advertising and promotion	2,000.00
Decorating	1,000.00
TOTAL	$40,900.00

Cash Flow Statement
Month of _____

Income		Expenses	
Wages	$_____	Rent or mortgage	$_____
Miscellaneous	$_____	Auto loan	$_____
		Gas & auto repairs	$_____
TOTAL	$_____	Auto insurance	$_____
		Life insurance	$_____
		Medical insurance	$_____
Savings	$_____	Homeowner's insurance	$_____
		Taxes	$_____
Credit Line	$_____	Loan payments	$_____
		Food: At home	$_____
		Food: Dining out	$_____
Home Equity	$_____	Telephone	$_____
		Utilities	$_____
		Household repairs, etc.	$_____
		Medical bills	$_____
		Credit card payments	$_____
		Interest expense	$_____
		Clothing/dry cleaning	$_____
		Travel	$_____
		Miscellaneous	$_____
		Savings	$_____
		TOTAL	$_____

Start-up Costs Worksheet

Furniture: Purchase price $_____

 Down payment required $_____

Fixtures: Purchase price _____

 Down payment required _____

Equipment: Purchase price _____

 Down payment required _____

Installation and delivery costs _____

Decorating & leasehold improvements _____

Deposits: Utilities and rent _____

Fees: Legal, accounting, consulting, etc. _____

Licenses & permits _____

Starting inventory _____

Supplies _____

Printing _____

Initial advertising & promotion _____

Miscellaneous _____

 Total Start-up Expense $_____

 Less Available Start-up Capital $_____

 Total Amount Needed $_____

Projected Expense Statement

Months	1st	2nd	3rd	4th	5th	6th
Rent						
Utilities						
Telephone						
Insurance						
Professional services						
Taxes & licenses						
Advertising						
Office supplies						
Office equipment						
Inventory						
Business auto expense						
Travel expense						
Entertainment						
Dues & subscriptions						
Salaries						
Owner's draw						
Loan payments						
Interest payments						
Miscellaneous						
TOTALS						

6

COMPUTERIZING YOUR OFFICE

In choosing your computer, the following criteria should be considered: the speed of the overall system and, related to this, the on-board (resident) software that drives the computer and its compatibility with other software packages that you may need down the road.

In addition, you'll want to check the system's upgrade capability (the ease with which you can upgrade components when the manufacturer offers new or revised product lines); and the system's overall compatibility with hardware offered by competing manufacturers.

Once you've decided on your business applications and software needs, the next step is buying the hardware. Hardware comprises not only the computer (central processing unit or CPU, monitor, mouse, and keyboard), but, peripherals (printers, modems, etc.) and add-ons (such as fax and graphics boards).

In a constantly changing computer market, your best bets include reading the appropriate computer journals, calling manufacturers' customer service departments, and—most important—comparison shopping.

Hardware prices can vary widely from store to store, so it literally pays to shop around. Talk to a variety of salespeople, but don't let them talk you into a purchase. It's your decision, and it's

too important to let it be dictated by a commission-hungry pitch-man. Take your time, then base your decision first on need, then on price. A low price is no bargain if you walk away with a computer whose memory or other functions are inadequate or—worse yet—incompatible with your software needs.

In selecting hardware, your main considerations center on

1. Current and future business applications

2. Software requirements

3. Necessary peripherals

4. Price

Term Warfare

The first challenge in hunting for hardware is surviving the initial cross fire of computer-speak. If you can master a few simple terms (and can type), you will likely master the computerized home environment.

First, a few basics. The core of a computer system is the central processing unit, which among other things interacts with software. Virtually all CPUs contain at least one floppy-disk drive, designed for either 5 $\frac{1}{4}$ - or 3.5-inch magnetic disks on which text, graphics, and additional data are stored. The main storage area, however, is the system's internal or "fixed" drive, also known as the hard disk, which offers increased storage capacity.

Storage capacity refers to the amount of data that can be saved and is measured in terms of bytes or characters (each alphabetic letter, number, or punctuation mark represents a char-acter). Under one million characters, floppy-disk storage is meas-ured in kilobytes, signified with a capital K: An 800K disk thus can store 800,000 characters. Storage of one-million-plus characters is measure in megabytes, signified with a capital M or MB. A 30MB hard drive thus can store up to thirty million bytes of data.

In addition to storing data (broken down into files), floppy and hard drives store and load software, allowing immediate access to a variety of programs.

A low price is no bargain if you walk away with a computer whose memory or other functions are inadequate or incompatible with your software.

The second element in the computer system is the monitor, or screen, available either in monochrome (black and white) or color. If you will be doing a lot of graphics work, a color monitor is essential. A mouse or other pointer device is necessary for most of the applications you will be using. There is also the keyboard, with a layout similar to that of a typewriter. Placement of function keys and the touch or feel of the keyboard are the major differences between models.

Next, you must consider peripherals—most commonly the printer and modem. Printers come in three forms: dot matrix, ink jet, and laser.

Modems transmit and receive computer data over telephone lines, connecting you electronically with the outside world—whether a print shop down the street or a client in Europe.

Modems usually are categorized by their transmission (or baud) rate, measured in bits (characters) per second (bps). Baud rates are typically 1200, 2400, and 9600, but more commonly range upwards of 14400. Modems of 2400 baud and higher are recommended for long documents. The higher the baud rate, the faster the transmission time and thus the lower your phone bill, since transmission time is billed as telephone time. In addition, there is the fax modem, which turns your telephone into a data retrieval/transmission center.

Finding the Right System

The leading-edge computer market is divided into two segments: the Apple Macintosh and IBM compatibles, the latter now dominated by Windows—a graphical user interface similar to that of the Macintosh. Here's where things get both interesting and a little confusing. The growth of desktop publishing has resulted in a flood of products, and it can be difficult to make your equipment choices. The choice does, however, essentially boil down to two

general computer environments: the Apple Macintosh or Microsoft Windows for the IBM PC and its clones.

What makes one better or worse than the other? That decision must be left in the hands of consumers. Each camp has its loyal followers. To set your business up, you'll need to decide what system you personally like best. Here is some general comparative information about each system.

	Macintosh	Windows
Ease of Use	Slightly better	
Ease of Learning New Software	Slightly better	
Graphics Capabilities	Comparable	Comparable
Price of Hardware		Considerably lower
Availability of Hardware		Slightly better
Availability and Cost of Service		Considerably better
Price of Software	Comparable	Comparable
Availability of Quality Software	Comparable	Comparable
Price of Peripherals	Comparable	Comparable
Availability of Peripherals	Comparable	Comparable
Compatibility With Other Businesses	Depends on market segment	Depends on market segment

Instead of outlining all of the available hardware (something that would fill the heftiest of computer magazines), we'll concentrate on the variables that you should consider prior to making a purchase decision.

Keep in mind, however, that you don't have to buy new. You can purchase good late-model systems through computer rental

stores (leasing with an option to buy), as well as through used computer outlets and private parties.

Computer Shopping Checklist

◆ Will the computer handle the software programs you've selected, as well as the programs you eventually may need?

◆ How much memory do the software programs require, and does the system need to be expanded to handle them?

◆ What will it cost to expand the system, if necessary?

◆ Is the storage capacity of the hard drive adequate?

◆ How much setup space is required for the computer, printer, and other equipment?

◆ Is the operating system user-friendly, with easy-to-use commands and a comprehensive instruction manual?

◆ What assistance or technical support does the hardware manufacturer provide?

◆ Is the monitor an adequate size and resolution for your needs?

◆ What does the basic system price include (monitor, additional software, etc.)?

◆ What are the terms and length of the warranty, and to what extent is labor included?

◆ What kind of supplemental service contract is available? What will it cost on a monthly or annual basis?

◆ Does the dealer offer ongoing support?

◆ What is the quality of printer output?

◆ How fast and loud is the printer?

♦ Is the dealer willing to put a quote in writing for comparison shopping?

♦ Is financing available through the manufacturer or dealer? At what interest rate?

♦ Is a leasing program available? Can you purchase the equipment at the end of the term? At what price?

Printers

Here we offer a quick roundup of the leading laser, dot-matrix, and ink-jet printers. Shop around for the best deal.

Laser. Laser printers work on a principle similar to photo-copiers. The image from the document stored in your computer is transferred to a drum by a laser beam. The drum then lays an electrostatic charge on the paper that attracts toner into the areas to be printed. The images are set in a series of minute dots that look like solid lines to the naked eye. Printer resolution is measured in dots-per-inch (dpi). The higher the resolution the sharper the image. Laser printer resolution is typically 300, 600, 1000, or 1200 dpi. Printers with higher resolution will generally cost more and print more slowly than their lower dpi counterparts.

Laser printers have become increasingly affordable in recent years, turning the home computer into a full-scale type and graphics center. You can create camera-ready copy for a print shop or generate your own finished advertising materials. Consider wear and tear on your equipment when deciding whether to do it yourself or to use outside services for reproduction.

Dot matrix These printers produce characters composed of hundreds of dots and are popular choices for the budget-conscious who don't require high-resolution printing capabilities.

Ink jet These printers produce pages of text or graphics by spraying fine jets of ink on the paper to form characters made up of many dozens of dark black dots. Ink-jet output is often confused with that of low-resolution laser printers. For this reason an ink-jet printer is a good compromise between the high resolution and high cost of lasers and the rough low resolution of dot-matrix printers.

Base your hardware purchases on current and future applications, needed software, available peripherals, and, of course, price.

Laptops and Notebooks

These lightweight, battery-powered computers can be used as standalone systems or as add-ons to your home computer, in both cases providing system access at remote locations—from the back seat of a limo or from a 747 at 30,000 feet.

CD-ROMs

This stands for compact disc read-only memory. It is important to remember that CD-ROMs do not replace hard drives and other storage media because you cannot save your own files to them. (Rewritable CD systems are available, but they are extremely expensive.) Instead, they are used as a means of storing huge amounts of ready-made data, including software applications. They are capable of storing 150,000 pages of text per CD as well as complex graphics and sounds. For this reason they are used to store everything from downloadable fonts to 60-million-word dictionaries to encyclopedias that allow users to listen and view as well as read. CD-ROMs are standard equipment on a number of systems and are now available at an increasingly reasonable cost.

Additional Accessories

Surge suppressors prevent damage to your equipment when there is a fluctuation in voltage. Basically a long strip or box with several power outlets and an on/off switch, they are well worth the minimal investment of less than $50, especially if you live in an area where power spikes or surges are common. You should note, however, that a surge suppressor will not protect unsaved data in the event of a power failure. For that you need an uninterruptible power supply (UPS). You should also be wary of power strips that are not surge protectors. They often look similar,

but their only function is to increase the number of outlets. Some surge suppressors come with guarantees that reimburse you for equipment that is damaged by power surges while you are using the suppressor.

Computer and printer covers protect your equipment from dust, which can sneak into the disk drives and under the keys to create problems. They cost an average of $10 to $20 each.

Screen filters reduce glare from windows or lights that reflect off the computer screen and into your eyes, causing eyestrain and headaches. Although a lower level of light is required for computer input than when working with a typewriter or writing by hand, it's sometimes impossible to prevent glare from a window. For less than $25, you can buy a mesh or plastic screen filter that will fit over the screen.

Disk storage boxes come in a variety of styles and sizes. Generally made of plastic, they not only allow you to keep your library of disks in one convenient place, they also help you by protecting the disks from electromagnetic fields, which can alter and even erase data from disks. You do this by keeping your disks in the box and keeping the box at least three feet away from electronic equipment (such as telephones, calculators, TVs, and stereos).

Shop Carefully

Again, we can't stress enough that it pays to shop carefully. Computer systems can range from below $1,000 to well over $5,000, with many of the lower-priced systems offering all the power and speed you'll need. It's very easy to overspend on technology that you won't really require and is available at a lower price on less expensive models.

Read the leading hardware and software computer magazines and comparison shop at a variety of stores. Also, as mentioned earlier, there's a good chance you'll be able to find a perfectly good previously owned system at a very reasonable price, particularly in light of recent hardware upgrades inducing users to trade up. Check the classifieds and local computer users groups for good buying opportunities.

Mail Order

The same principle applies with mail-order computer houses that advertise in computer magazines. It is possible to save hundreds of dollars buying brand-new systems from mail-order outlets, and you can also buy components.

In either case do some preliminary research before buying. To the extent possible, know what you're getting.

Financing and Leasing

Many of the major computer manufacturers offer their own financing packages, and major computer hardware stores do as well. Of course, monthly payments are required and, in some instances, interest rates may be above those you could obtain through a personal or business loan. Here, too, it pays to comparison shop to be sure you get the best financing available.

Leasing can be an advantage, since it offers the opportunity to trade in one computer on an upgraded model down the line. You may also have the option of buying the hardware at the end of the term for a preset price. Leasing is discussed in more detail at the end of this chapter.

Back-up Service and Support

As with any electronic device, computers break down and need to be repaired. This can be costly, not only for parts and service but in lost production time. Therefore, be sure any new system you're planning to buy has a good warranty and has a service contract available. (In some cases, used computer stores also offer limited warranties, so be sure to ask when shopping the market.)

New system warranties vary greatly in duration. Any warranty or service contract should cover both parts and labor. One drawback, no matter how good the warranty, is that you'll be required to ship the computer to the service center, perhaps at your own expense. If possible, try to have the dealer sign a statement covering the cost of shipping if the system needs service within the warranty period. Also, ask whether loaners are provided while your system is being repaired.

Service or maintenance contracts may seem like a heavy expense when you've just laid out hundreds and perhaps thousands of dollars on a new system. However, in the event that you need to have major work done—replacing a disk drive, for example—you'll be thankful you made the investment.

Equipment Buying: New or Used?

You can buy new or used equipment from dealers and independent sellers. Or, if you don't mind rummaging through other people's lives, you can find incredible bargains on supplies and equipment at garage sales, swap meets, thrift shops run by charitable organizations, or auctions. Often, it is well worth the time involved because the savings can be significant.

Before investing money, do your homework. Talk with other business owners to find out which brand of equipment they use and why they prefer it over other choices on the market. This will help you make the best decision based on your needs and budget.

Also, talk with independent dealers who carry a broad line of similar types of equipment. They can give you insights on maintenance, longevity, service contracts, and prices.

They will also be able to tell you when to expect a drop in price for the item you are interested in, although you can count on seeing sales on major equipment such as computers and other big-ticket items at store openings and during special sales.

Cash or Credit?

Unless you are planning to pay cash for an equipment purchase, you will either make a down payment and set up a payment schedule with the dealer or take out a loan with the bank. With interest rates for some bank loans currently running at about 10 to 12 percent, you might be able to find a dealer who is willing to handle the financing at a lower rate.

Although this method is a less effective way of establishing your credit, it will serve to get the business going. Another advantage is that if you should find yourself in a tight cash-flow situation at any time, a private dealer/lender with whom you have

a personal relationship is much more likely to be flexible...without charging you a penalty for a late payment.

Looking for Bargains

There are a number of ways to save money buying equipment and supplies, generally through third-party vendors and mail-order sources. Apple and IBM equipment is sold only through authorized dealers, while IBM clones and third-party peripherals are available everywhere from small stores to mail-order businesses run by electronics whizzes who make their own computers and accessories.

A word of caution here: You need reliable equipment that is backed up by warranty. Whatever your decision on brand, make sure you are protected with a company that stands behind what it sells.

Shopping around is the only way to get the best deal on equipment and software. If you live in a small town, buy a newspaper for the nearest big city and look for computer ads.

Buying Used Equipment

When buying used equipment from a dealer, the chances of it being in working condition are generally pretty good. Dealers have a reputation to uphold and will stand behind their merchandise, especially if they are firmly established in the community. There are, however, several ways you can scout around to ensure that the one you decide to work with is reputable.

◆ Find a store with membership in the local chamber of commerce. Although this is not an iron-clad guarantee, it does indicate the store owner's interest in following business standards established within the community.

◆ If the dealer also sells new equipment, it is highly likely that much of the used equipment has been traded in by people who are upgrading. You may be able to get an excellent deal on an item that has been carefully maintained by the former owner.

♦ If there is a service department on the premises, you can be assured that used equipment has been reconditioned before being put on sale. It also increases your chances of getting fast, inexpensive, and reliable service if needed at a later date.

♦ Used equipment dealers realize that it is not always easy to find a buyer and should be willing to work with you. Shopping around will give you an idea of average prices and will give you the information you need to negotiate. If the dealer won't work with you, keep looking.

♦ Know the terms of the warranty. Even used equipment should be covered for a short time for parts and labor, especially if it has been overhauled or reconditioned.

♦ Ask the dealer if he or she is willing to agree to offer a trade-in allowance on the item you are buying when you decide to upgrade. Of course there will be stipulations based on wear and tear and time; however, the dealer should be willing to consider it.

♦ Contact your local Better Business Bureau to see if any complaints have been made about the dealer.

Buyer Beware

There are other ways to find and buy used equipment, but it falls into the realm of "buyer beware." If you choose to deal with private parties through the classified ads or with auctioneers selling off the inventory of a bankrupt business, you must be willing to take a chance. Although the prices will be easy on your budget, the cost of repairing a malfunctioning machine could result in a long-term drain on your profits.

This is not to say that there aren't good opportunities out there. The secret is to know what you are looking for and to have a good idea of how it should work. Do your research by visiting with reliable used-equipment dealers before you start looking.

Ask a private party how long they have owned the equipment, why it is being sold, and if there are any repair bills that you can look at. Trust your intuition in this kind of situation; if the

party seems truly interested in providing you with as much information as possible, chances are the equipment is as claimed. If you're told it needs a new part, for example, find out what the replacement part costs and ask if that amount could be deducted from the selling price.

Ask a private party how long they have owned the equipment, why it is being sold, and if there are any repair bills that you can look at.

When you go to see the equipment, make sure it runs. Test it out, if possible, and measure it to make sure it will fit into the space you have available. Also, see whether you can move it yourself or will have to arrange transportation.

Before buying the equipment try to find out if there is a servicing outlet nearby or if you must send away for parts, which can be time-consuming and costly, especially when it comes to older models.

Occasionally you will run across an individual who is making payments on a piece of equipment still under warranty. This can be a great deal for you since he or she will probably accept a small amount of cash and let you take over the payments. Be sure to transfer ownership in writing to prevent complications if you need service under the warranty.

Auctions

Auctions are an excellent way to get equipment and supplies at good prices. Watch the business and classified sections of the newspaper for ads about upcoming auctions. The ad will include dates, the reason for the auction (liquidation or bankruptcy), location, time, and a partial list of items. In most cases there will be a preview, enabling potential buyers to view the merchandise before the bidding begins. Take advantage of the preview to inspect and select the equipment you want to bid on. A fee will be required for a bidder registration number, which is held up when

you make a bid so the auctioneer's spotters know who has purchased a particular item.

There are two rules to remember at auctions:

1. Cash or a personal check for the full purchase amount must be paid on the day of the auction.

2. Don't move your hands or make significant gestures during bidding or you might find that you have purchased something you didn't want.

Leasing Equipment

Leasing is defined as a long-term agreement between two parties for the use of a specific item. The person who leases is known as the lessee, while the owner of the item is referred to as the lessor.

Despite the fact that you do not own the equipment when you lease, and so cannot take advantage of depreciation on it for tax purposes, there are still many benefits for the beginning business owner.

Leasing lets you try out a piece of equipment for a given period of time to determine if it is the best product for your needs. Although you are locked in to the terms of the lease, most lessors are flexible.

Know What You Need

Of course, the best way to enter into a lease is to be absolutely clear about what you expect the equipment to do for you. The service representative from the leasing company is well-versed in tailoring equipment to the customer, so do not hesitate to ask questions about capabilities.

Most lessors offer good maintenance contracts, as they want to protect their equipment. Check to see what parts and/or labor are covered before signing the lease. The lessor should also be willing to provide technical advice at no charge, may be willing to offer installation and setup of the equipment, and may also provide training, if required.

Payments can be arranged to fit your budgetary needs on a monthly, semiannual, or annual basis. This gives you the freedom

to schedule payments for peak cash-flow periods. You can also negotiate the rates and length of the lease to keep monthly operating expenses at a minimum.

Conditional Sales Agreement

Under the provisions of a conditional sales agreement, you become the owner of the leased goods from the agreement date. At the end of the lease period, you are required to purchase the item for a preestablished price. This is often referred to as a balloon payment and should be agreed upon by you and the lessor prior to your signing the lease.

The conditional sales agreement, unlike most other leasing contracts, gives you the tax advantage of claiming depreciation on equipment. Depreciation refers to the decrease in the value of an asset because of wear and tear over a period of time.

You are entitled to deduct depreciation from your income tax based on the value when new, the estimated life of the item, and the value at the end of that estimated life. It is best to work with your accountant on determining depreciation of fixed assets.

There is seldom a down payment, other than the first month's lease amount, required on a leased item, since leasing is generally 100 percent financed for the terms of the agreement. This frees your start-up or working capital for other uses.

True Lease Agreements

You can write off lease payments on your income tax, but only if you have a true lease contract. Under a true lease, the lessor owns the equipment at all times during the contract period. If you decided, at the end of the lease, that you wanted to buy the equipment, you would have to pay whatever purchase price was decided by the lessor.

Financial Lease

The financial lease covers a period that does not extend beyond the estimated life of the equipment. Payments must be made as stipulated on the date due and made consistently to the end of the

lease. It usually puts the responsibility and cost of maintenance on the lessee.

Operating Lease

The operating lease generally requires the lessor to handle maintenance of the equipment. It offers the option of canceling the lease, but only if a cancellation clause has been included at the negotiating stage.

The most important aspects of leasing are the terms outlined in the formal lease. Have the lessor draw up a proposal for you, based on everything you have discussed in an initial meeting. If you have any trouble understanding the terms of the proposal, have your accountant review it with you. In fact, it is a definite advantage to have the final lease agreement checked by your attorney or accountant before you sign it.

Your Lease Agreement

Your lease agreement should include

- ◆ The length of the contract in months or years
- ◆ The rate you are to be charged, which is usually a percentage of the total purchase price computed on a monthly rate
- ◆ Your payment schedule
- ◆ A purchase option, if applicable, at the end of the lease
- ◆ A renewal option, if applicable, which allows you to carry the lease over for an additional period of time
- ◆ A cancellation agreement in the event you want or need to cancel the lease
- ◆ Maintenance stipulations (who pays for parts and labor)
- ◆ Substitution options if updated equipment is introduced and you want to take advantage of improvements
- ◆ Any provisions particular to the lease, including tax allowances for depreciation, insurance liability in case of loss or damage, and your responsibilities in reporting a move or other major change

Whether you decide to borrow, buy new equipment, find good used equipment, or lease, be sure to get exactly what you need to keep costs to a minimum. This is especially important during the early stages of your business when cash is bound to be tight. You can always upgrade or add to your equipment inventory as profits increase.

Notes

Key Points

Personal Thoughts

Additional Research

$1,29.99 99

129.99

43 32 51

AutoZone

5.47 28.94

4.47 oil 20.94

1.49

6.97 30.38

5.47 22.99

7

POTENTIAL CLIENTS

Your biggest order of business is attracting customers. How do you get their attention? Basically, by advertising, maintaining high visibility, and generating positive word-of-mouth referrals.

High visibility is partially a matter of your physical plant (Does your store draw the eye? Is your sign prominent and attractive?) Nonadvertising methods, such as posting handbills, putting flyers on cars in parking lots, and so on, can also help you maintain high visibility.

Who will buy specialty coffees and teas? People who like the good things in life and can afford to pay for them. People who buy gourmet coffees are willing to pay more for quality.

The world's finest coffees, selling for $6 to $10 per pound actually cost an average of $0.10 a cup. Compare that cost per cup to such beverages as wine or soda, and you can see that even the most elite and expensive coffees are a flavorful bargain. This will also give you cause to realize that the $0.75 or more that you pay for a cup of standard restaurant coffee, often bitter from sitting in an urn all day, is highway robbery.

While gourmet coffees and teas enjoy an upscale image, it is up to you to convince your customers that they are indeed getting a bargain. You will want to offer them the best prices.

If you are planning to open a store, you will want to have a grand opening and invite everyone you know. Make up flyers to pass out to every store in the area. Place an ad in a local newspaper and advertise a special grand opening offer. (For example, purchase one pound of coffee and get another half pound free.) Your grand opening must be splashy. You will want to invite the people who you believe will provide your greatest repeat business.

Just plain "impulse" leads many people into a coffee and tea store. They get a sudden inclination to stop in for a cup of freshly brewed coffee when they pass by and smell that irresistible aroma floating through the air.

*If you have a job without aggravations,
you do not have a job.*

Malcolm S. Forbes

Ground Floor Marketing

Make a list of everyone you know. Even if the names on your list seem unrealistic as far as potential clients go, keep in mind that just about everybody loves a cup of hot coffee or tea. Your list will help you plan your business approach, develop your sales skills, and recruit new customers.

Start by including the names of the following people. Undoubtedly you will think of other contacts that are not outlined below, but this will give you a strong starting point.

◆ Relatives and friends who live in the area

◆ Acquaintances from work, school, church, or social groups

◆ Local retailers you do business with on a regular basis

◆ Members of clubs and organizations in the area

◆ Acquaintances through sports or hobbies

◆ Your doctor, dentist, banker, and other service providers

♦ Professional people you know, including real estate brokers, insurance agents, and corporate employees

There probably isn't a person among them who doesn't get the urge for a steaming hot cup of strong coffee or a soothing cup of tea at times. By letting them know that you are in business, the seed will be firmly planted and the next time that they get the urge, they will undoubtedly come to you.

As soon as you are ready to open the shop (or your pushcart) give out business cards and brochures and ask everyone to pass the word along!

Potential Customers

Friends and Relatives

Acquaintances and Service Providers

Business Contacts

8

A MARKETING OVERVIEW

Now that you have selected the kind of business you want to own, it is important to explore the need for it. A process known as *marketing research* will provide you with the information you need to develop your business, plan methods of distribution or promotion, and set prices that are tailored to the audience you hope to attract.

In addition, your marketing research will provide you with information that will help when you make decisions about a location, hours of operation, the specific types of services and/or products to sell, and how to gear your advertising.

Identifying Your Market

The process of identifying your audience may seem to be an extremely complex process; however, you can develop a perfectly workable and valuable marketing report by using the following guidelines and adapting them to your particular situation. Basically, there are five factors used to target the market.

Five Factors Used to Target Your Market

1. Population
2. Income
3. Competition
4. Product or service market match
5. Desire

1. ***Population.*** The number of households in the region you
 are considering for your business is crucial, as you must
 have a sufficient population base to produce the sales you
 need to generate a profit. Equally important is the circu-
 lation and age range of readers of the publications where
 you will be placing your advertising. If, for example, you
 are planning to sell products for infants, a publication
 whose readership is largely of retirement age would be
 inappropriate. It would, however, work in your favor if
 you were promoting health products or gift items.

2. ***Income.*** Your potential customers must have the income
 to purchase goods and services. Consumers in the thirty-
 five to sixty-five age group generally have considerable
 income to spend on household items, personal grooming,
 and sporting goods. This is not to discount the over-sixty-
 five age group, a large and growing segment of the
 nation's population which, depending on the region, has
 adequate discretionary income (money after taxes and
 necessities) to spend, or the eighteen to thirty-five age
 group, which is a desirable market for clothing, personal,
 and recreational items.

3. ***Competition.*** Competition shouldn't be a negative fac-
 tor. Rather, it should spur you on to stretch your creativ-
 ity by creating something brand new or creating a similar
 product or service that is superior to those being offered
 by the competition—through quality, selection, or price.

4. ***Product or service market match.*** Basically, this
 means that you must be able to attract those consumers

whom you have the resources to serve. For example, if your idea of the perfect business involves national distribution of your patented weight-training equipment, you must reach an audience that is receptive and interested in bodybuilding through a carefully designed advertising campaign, and have the financing available to supply and ship the product.

5. ***Desire.*** Your objective is to match your product or service to the needs and desires of a particular group of consumers who will be responsive. It is often difficult to figure out exactly what your target market wants. However, through observation of what the competition is doing, it should be possible to recognize a need.

Market Research Techniques

Large corporations often have in-house marketing staffs who conduct extensive research on a continuing basis to ensure that the products or services being offered are in line with the marketplace. Obviously, this is an expensive and time-consuming process; one that you undoubtedly want to avoid.

Through several easy and inexpensive methods, you can find out everything you want to know about your potential market. But first you need to determine exactly what information you need. It might be population or economic trends, the number of new homes built locally within the last five years, or how many people are active in a particular hobby.

The greater the difficulty, the more glory there is in surmounting it. Skillful pilots gain their reputations from storms and tempests.

Anonymous

The nearest Census Bureau office and your local chamber of commerce are consistently good sources for regional statistics. The

reference librarian at the public library can steer you toward other data that will give you the specifics you seek.

The Small Business Administration (SBA) compiles extensive marketing information, in addition to material on operating procedures for specific types of businesses.

Check the *Encyclopedia of Associations* at the library to find the names and addresses of various trade associations for the coffee and tea industry. Trade boards can provide associates with marketing statistics, management tips, and a wealth of valuable information. Often it only takes a phone call to get more details than you could ever use.

Another excellent source of information on population, income, and sales figures is the annual survey of buying power published by *Sales and Marketing Management Magazine,* which breaks the information down by county and cities in the United States and should be available through the library.

Media Kits and Personal Contact

The advertising departments of local magazines and newspapers undoubtedly have a "media kit" available for potential advertisers, which they will gladly send you upon request. These packets contain a breakdown of their advertising rates and specifications, a description of why advertising with them is to your benefit and, most important, a profile of their readership. A friendly conversation with one of their salespeople should give you a wealth of data.

The merchants from whom you will be buying supplies, equipment, and products are another excellent source. They can give you a good rundown on trends, as well as an overview of current sales figures for their products. Since they are hoping you will eventually use them as a supplier for your business, they will be more than happy to give you free information.

It is, of course, often possible to gauge what the competition is doing and to glean information from them. There are two approaches when talking to people who are soon to be in direct competition. One is to be up-front and honest about your business plans and appeal to their sense of "industry spirit."

Surprisingly, you will find the direct approach works in the majority of cases as most people are genuinely interested in and

supportive of others trying to make it in their field. It is better for everyone if "industry" standards are maintained and competitors have a healthy rapport. And, except in extreme situations such as a very small community, there is generally enough business to go around. It shouldn't be difficult to capture your share of the market, especially if you can develop something unique to attract them.

On the other hand, if competitors are less than receptive, it may be necessary to do some detective work to get the information you want.

Focus Groups

If you really want an in-depth marketing study, consider gathering together a group of people (family members, a social or church group, or friends) for a focus session to determine if your product or service will match the needs of the prospective audience. This involves presenting your proposed business idea, with product samples if applicable, and creating a questionnaire that calls for specific answers from the group members.

Referral Discounts

Here's a way to boost your income that will make your clients happy to promote the business. Offer clients who refer a new customer a discount on their next purchase.

This method is often used by companies when they are testing new products and, in fact, there are private companies that do nothing but put focus groups together and set up testing sites in stores, shopping malls, and street corners to obtain spontaneous and objective input from potential consumers.

The questions you would want to include on your question-naire would cover such aspects as how often members of the focus group have frequented a similar store in the past, what they liked about it, what they found to be unsatisfactory, how they feel it could have been improved, whether they would be willing to try

another, and vital statistics such as their age, income, and any specifics that relate to your proposed business.

With a coffee and tea store, the questions might include a breakdown of the top three reasons people in your focus group would consider frequenting your store, what they feel they would be most likely to purchase, how much they would be willing to pay, and what else they might want to find offered in the store. This kind of information will give you an immediate edge on the competition when you are ready to start advertising.

Analyzing Your Research

The bottom line in conducting your research is that you want to zero in on information that provides insights on the potential for your business idea before you invest time, money, and energy in setting it up.

If, for example, you were considering starting your coffee and tea store in a small community, and your focus group information indicated that only 10 percent of the local residents would consider using your service or product, you would definitely want to reconsider the validity of your concept or figure ways to promote it on a broader scale.

On the other hand, if your marketing research pointed out that 90 percent of the population thought it was a great idea and that 50 percent would have an immediate need for it, the potential for your business would be much greater, and proceeding with the idea would probably guarantee a profitable venture.

Buy an inexpensive notebook to help you keep track of your marketing data. Use a separate page for each category you are researching. The notebook will serve as your personal, ongoing market study to be reviewed and amended as your business grows and the audience you are serving changes.

Plan to update your information as new studies are published (generally an annual event), indicating changes in population, economy, or buying and spending trends. Most newspapers publish synopses of local, state, and federal studies of this nature, so maintain your notebook as appropriate. You should also reserve several pages to record comments and suggestions from customers once your business is established, which will help you personalize

the service to the market and keep you a step ahead of the competition.

Spend as much time as needed to feel comfortable about your marketing project. The important point is that the results of your research are comprehensive enough to provide you with concrete information on who your potential customers are and how you can best reach them.

Review

I have completed my entrepreneurial profile to determine my strengths and weaknesses. _____

My friends and family have given me additional input based on the profile. _____

I am aware of the advantages and disadvantages of going into business for myself. _____

Time is not a problem; I can easily devote the time I'll need to build my business. _____

The important people in my life are supportive of my decision. _____

I have analyzed my cash flow to ensure that I can support myself and my family for at least six months or until the business is solvent. _____

I feel confident about my future as a business owner at this point. _____

I know what people want as far as my business is concerned. _____

I have conducted informal studies to determine my potential customers and understand their needs. _____

I have analyzed the competition, know what they offer, and have a general idea about their success ratio. _____

I have done my marketing research and know how to get in touch with the audience I want to reach. _____

I have contacted the trade association for my industry and have accumulated facts and figures regarding the pros and cons of starting my own business. _____

I feel that my product or service is marketable. _____

Notes

Target Audience

Publications to Reach Audience

Additional Research Information

9

CHOOSING A SITE

What if you found the perfect site for your coffee and tea store, bought equipment and supplies, moved in, decorated the premises, opened your doors, and then had no customers? The problem is probably that you are located in an area where no one is interested in specialty coffees and teas. It's also safe to say that you didn't research your market, or at least you did not do so thoroughly.

To avoid the above scenario, do some marketing research. This will provide you with the information you need to

♦ Select the proper location

♦ Identify your best hours of operation

♦ Purchase the correct products to sell

♦ Gear your advertising to potential customers

The preceding chapter, "A Marketing Overview" told you how to do this. The time you spend on market research will definitely be worth your while.

After you have completed your marketing study and determined the best area for your shop, explore it to determine the best choices in your price range. Since gourmet coffees and teas are generally purchased by upwardly mobile middle-class buyers, you

will be more successful if you can locate in or near a neighborhood where these people live.

This may cost you more rent than you were planning to spend, but it may be worth it to identify other expenses to cut in order to offset this critical outlay. Besides, because you are located in a fairly well-to-do area, the likelihood of your success is enhanced by the presence of a market that has the money to keep you going.

Guidelines for Commercial Locations

You want space in an area with commercial appeal, including easy access, adequate parking facilities for customers, and good visibility for your sign or window treatment.

A storefront on a main commercial thoroughfare is definitely desirable, but not absolutely necessary if you can find space in a busy office building or convenient business park, for example. You will want to make sure there is no other coffee and tea shop nearby.

Usually a landlord charges on a per-square-foot basis, so you need to determine the amount of space you need to accommodate your equipment and your need for work and customer space.

In considering your space, remember to include storage, an area that is pleasant for customers, as well as ample work space for you and/or an employee. You don't want to waste space, but you might want to consider growth demands down the line.

Selecting the right location is vital to the success of your business. When looking for a commercial location the foremost considerations are

1. The community you want to live and/or work in, based on family needs, finances, your preference for a particular area because of health reasons, or the fact that you have an established reputation in a certain area

2. The locations available within that community

These factors are interrelated. You may want to settle down in an area with a limited number of suitable business locations available. Or you may have run across a number of viable sites in several

communities or locations, in which case an investigation of each must be conducted, covering each of the following points:

1. ***The type of business you are planning to operate.*** Remember that retail, wholesale, and service businesses have slightly different requirements as determined by the type of product or service being offered and the market potential in a specific area.

2. ***The demographics of the area.*** You should find out the number of consumers who want or need your product or service and are willing and able to pay your price, the median income and employment opportunities, the age ranges of the major population group, and the volume of retail trade and projected expansion data (available from census reports and chamber of commerce business reports accumulated during your market research project).

3. ***Competition.*** You must determine how many other similar establishments are serving the market and how their businesses are going to decide if there is room for your new venture. The best way to do this is by making a list of businesses from the phone book that you feel will be in direct competition and, if feasible, visit their locations at different times of day to observe the activity levels and to talk with employees, who should be willing to answer your questions if you approach them in a friendly way. Often, the same kind of research can be accomplished on the telephone.

4. ***Traffic patterns.*** Is your proposed location close to freeways, major intersections, and/or a central business district? Is there sufficient parking? Is the foot traffic past the location strong and steady enough to guarantee walk-in trade that will generate sales and profits? The ease with which customers can get to your location is a major consideration in terms of success.

5. ***Your image.*** Before you go scouting for locations, decide on the image you want to project, such as quality products at good prices, superior service, and convenience.

6. **The product or service.** Generally, there are specific areas within a marketplace that cater to consumers in specific income brackets and/or occupational groups, i.e., executives, blue-collar workers, students, etc. Consider your product/service and the projected number of potential buyers within the community.

7. **The amount of rent required.** Locations having the highest potential of profit through consumer traffic (such as a prime downtown site or shopping mall, a corner shop, or a store with good frontage) are more expensive because competition keeps rents up to the maximum. The trade-off is an increase in sales and a lower advertising budget because of the visibility factor.

Retail Businesses

The guidelines indicated above are applicable for retail businesses. Poor location is one of the chief causes of failure among retail stores, but, on the other hand, the right location can be all it takes for even a mediocre business to thrive and grow.

Wholesale, Manufacturing, or Mail-Order Businesses

Where you locate a wholesale business depends on your market. If dealing primarily with local retailers or customers, your location should be within easy driving distance of your clientele.

However, if most of your business is conducted through the mail or delivery services, you can base your selection on the best rent available and the convenience factor for you and your staff. When choosing a location for your wholesale business, warehousing needs are a vital consideration, as is projected expansion.

Before Signing a Lease

Unless you're planning to purchase the commercial location, rent your location from a family member or accept a temporary agreement in a location that is for sale. But keep in mind that you will be required to sign a lease before moving in. The most desirable agreement for you as a new business owner is a one- to two-year

lease with a renewal option and a guaranteed rate for rent increases over a five- to ten-year period.

Rent for a commercial location is established either on a flat rate or a percentage basis. Under the flat rate, rent is generally based on the square footage of the shop and on the location or, in some cases, on potential volume. The percentage base involves a base amount of rent plus a prearranged percentage of monthly sales.

Your lease will also cover a number of other points, such as the liabilities and responsibilities of the landlord and of you, the tenant; i.e., who is to pay for utilities, specific repairs, renovations, tax increases, etc.

The lease may contain stipulations about the size of the exterior sign you can erect, hours of operation, insurance coverage, and assignation of the lease to another party (a sublet).

Before signing a lease to set up your business, make sure that electrical lines are adequate enough to handle high-volume usage, that you have rest rooms for employees and/or clients, and that there are convenient parking areas. Also check with the leasing agent to be certain you can make leasehold improvements, e.g., storage shelves, at your discretion as the business warrants it.

It is recommended that you have an attorney review the lease carefully before you sign it to ensure you understand all of the clauses and to serve as a negotiator, if necessary.

The Right Place for Your Gourmet Store

There are two main factors to consider in choosing the right location for your shop: (1) Is there a lot of foot traffic? and (2) Is there enough space inside the shop to properly display your inventory?

The ideal location for a shop is in an area that has a steady flow of foot traffic; a large part of your business will come from walk-in trade. Storefronts in busy retail shopping areas are generally more expensive to rent but give you the audience you need for success.

Sharing Space

Another option is to talk with owners or managers of established stores—for example, a gourmet food shop—who might be interested in having you share their location. You are assured of a built-in customer base. The store owner benefits from an exciting addition to the business that may increase sales. Both you and the owner benefit from reduced rent.

Other owners might be willing to lease you space on a percentage-of-sales basis, which would keep your rent payments in line with gross receipts. If you do discuss this kind of arrangement, however, be sure you come up with a mutually satisfactory contract that stipulates a ceiling on each month's payment. In the event you start to earn a lot of money, you want to make sure you are not overpaying your landlord for the privilege of the location.

Setting Up Your Space

Take the time to figure out a floor plan for your coffee and tea shop so that you know exactly how much space you need to be comfortable and to display your wares. If you plan to sell only coffees and teas you will need much less space than if you decide to include such accessories as mugs, coffee grinders, home espresso makers, candies, and books.

Most coffee store owners told us that their profits were definitely enhanced by having additional products in the store. When we asked what one item they would add to their inventory to increase sales, many said greeting cards. Since many people give gourmet coffees and teas as gifts, having the convenience of greeting cards in the same store is a plus.

Leave room for a display of coffees and teas, perhaps some greeting cards, a coffee grinder, a scale, a cash register, a sink and refrigerator or freezer, as well as room to move around. Figure out how much space you need and then look for a store that has nothing less.

If you rent a tiny shop, you will want to make sure that there is space to keep your overstock either in the basement or some other nearby area. Some coffee gourmets have told us that coffee

beans are best kept in a freezer, so you may want to consider purchasing one for your storage space if there is no room for one in your store.

Creating Atmosphere

Think of creative ways to maximize shelf space and to make your store look inviting. Use the aroma of fresh coffee and the sound of some soothing and relaxing music to attract customers. Music will often pull people in and keep them browsing. Some gourmet shops play New Age or classical music (which they also sell) to provide an oasis-like atmosphere. Thus, your shop becomes an enjoyable place to stop and perhaps sip a flavorful brew before going back to the everyday world.

The typical coffee specialty store has rows of glass cases or large glass jars with aluminum or wooden lids where coffees and teas are displayed. There are shelves and counters full of coffee makers, grinders, and other related products. Many also sell gourmet food and cookware. Some are set up as cafes or delicatessens as well as coffee and tea stores.

In terms of atmosphere and decor, you can go with high-tech, ultramodern, or cozy. Most of the shops we visited opted for the latter and went with an antique, country-store style, for example, displaying old wooden coffee grinders and packaging their coffees and teas in brown bags. Be creative and make your shop special. There are many ways to make a store warm and inviting, and a lot of this atmosphere will be based on your personal style in decorating. Have fun dreaming up an environment that will keep your customers coming back for more.

The Right Place for a Coffee and Tea Pushcart

Pushcarts used to carry a "peddler" image in the days when rag dealers and knife sharpeners wheeled down dingy alleyways selling their wares. Nowadays, pushcarts are considered charming and innovative. Their simple accessibility as well as their nostalgic effect attract customers.

If you are considering selling gourmet coffee and tea from a cart, you should know that your best location is in a busy,

prosperous shopping mall. An average mall, depending on location, will lease a cart for $200 to $1,000 per week, plus 10 percent of weekly sales over a certain amount. Since your investment in selling coffees and teas this way is minimal, healthy rent will be your only large expenditure. Some malls rent carts to vendors. For others, you need to provide your own.

Purchasing your own cart is best because after your initial outlay, you do not have to pay a monthly rental fee. In addition, with your own cart, you can go anywhere you like. If you decide to sell your coffees and teas only during the summer, or during high-activity seasons like Christmas, and you find a busy mall willing to have you rent a spot only during those times, you could wind up with very handsome profits.

> *God gives every bird its food, but He does not throw it in the nest.*
>
> J.G. Holland

Setting Up Your Business at Home

If you have decided to start your coffee and tea business with a pushcart set up in a mall or on a main street that guarantees foot traffic, it is quite feasible to run your business from your home. There is a lot of appeal in operating your business from home. Thousands of successful businesses have been started in a basement, a spare room, or on the kitchen table. It's true that Henry Ford founded the Ford Motor Company in his garage and Jean Nidetch started Weight Watchers in her living room as a support group for friends who wanted to lose weight. Both of these businesses, and many more like them, became successful multi-million-dollar corporations despite humble beginnings.

The Small Business Administration estimates that there are close to twenty million home-based businesses in the United States today and, of these, more than 30 percent are owned and operated by women.

Starting a home-based business has provided an opportunity for many people who might otherwise never have the chance to become business owners. Women, especially, have discovered that they can build a profitable, satisfying business at home while still being available for their families.

For others, a home-based business is the ticket out of the world of the urban commuter. In fact, a home-based business is the perfect way to try something new to see how it works while still working another job to pay the bills. Once the business has proven itself and is realizing a profit, you can leave the job to devote all your time to your new venture.

Couples often find that investing time and energy in building a business together at home develops stronger relationships in addition to increasing joint income. For the retired and for those with minor physical disabilities, it is a path to staying involved, exploring self-sufficiency, and guaranteeing that there can be a profitable future.

Start-ups Never Change

As with any new business, whether located at home or in a commercial location, it is important to follow the basic guidelines for start-up, including conducting a market survey, drawing up a business plan, setting goals, reviewing capital needs and projected income, developing an advertising campaign, and establishing a professional image.

Review

If setting up my business at home, I have

Checked with the city and county offices in my area regarding required licenses and permits and zoning regulations for the home office _____

Set aside a room or an area in my home that will be used exclusively for my business _____

Checked prices on storage units, worktables, or shelves _____

Planned a layout that I feel will work well for the location and my storage, display, and office needs _____

Installed a separate telephone line and purchased an answering machine, or contracted with an answering service _____

Set up a separate business bank account _____

Informed friends and family of my business routine and specific working hours to reduce interruptions and distractions _____

If setting up in a commercial location, I have

Investigated rental rates for the area I am interested in leasing _____

Checked traffic flow, parking, and foot traffic around my proposed location _____

Determined that my business is compatible with others in the area _____

Talked with my prospective landlord about improvements, maintenance, and rent increases _____

Had my lawyer check the rental agreement and any local zoning regulations _____

Secured the necessary operating licenses and permits from the city and/or county _____

Notes

Key Points

Possible Locations

Size and Floor Plan Considerations

10

PROFESSIONAL BACKUP

\mathbf{F}rom the start-up stage and as your business continues to grow and prosper, you will need the assistance of several professionals, including a lawyer, an accountant, and an insurance agent.

The best way to find a professional is, according to the majority of business owners, through personal recommendations from other entrepreneurs, especially those in similar businesses as yours, and from friends or relatives. The most important factor is that the person doing the recommending understand exactly what you will need from the professional you will be hiring.

For example, your cousin's divorce lawyer is probably not as well suited to helping you draw up a partnership agreement as the attorney a friend used for help with incorporating his business.

Before making a decision, talk to several recommended professionals until you find someone who can best satisfy your needs for the business as outlined below and who has a fee structure you can afford.

Equally important: Your choice should be someone you feel comfortable with, especially during periods when you find yourself having to call several times a week to resolve a problem or complete a specific task.

In many cases, because attorneys and accountants often work on a particular business matter in conjunction with one

another, the attorney you select may be able to suggest an account-
ant who can properly service your business, or vice versa.

If you're planning to hire an attorney or an accountant, you
should start interviewing likely candidates eight to nine months
prior to the date you plan to start the business. This will give you
time to find a suitable match and give the attorney time to take
care of all pre-start-up functions, such as assisting in the structure
of your business or helping with your business plan.

Professional Services

Legal

An attorney with broad-based expertise in business can help you
with such matters as raising capital; the legal and tax ramifications
and benefits of various business forms including sole proprietor-
ship, partnership, or corporation; name clearance to ensure that
you are not using a name already designated by another company;
legal tips on operating in your desired location; and help you file
all necessary legal papers and documents needed for financing,
establishing your business form, and so on.

He or she will review contracts and lease agreements and can
provide support with collection problems. The lawyer you select
should also be willing and able to represent you in the event of
any claims that are brought against you or lawsuits you initiate.

Fees

Depending on your attorney's expertise, reputation, and the mar-
ket he or she serves (metropolitan area versus small town, for
example), fees will differ dramatically.

In a small community, lawyers often charge a set rate for the
job being done, while "city" lawyers typically charge by the hour
with fees ranging anywhere from $65 to $250 per hour.

This does not include the extraneous expenses involved, such
as the $300 to $1,000 cost of incorporating (depending on the
state you operate in). Neither do these fees include supplemental
costs, such as travel and telephone, incurred by your attorney in
the handling of your case.

A good way to determine the range of fees in your area is to check with your local chamber of commerce or the state bar association, generally located in the capital city. The bar association may also be able to provide you with information about a particular attorney's reputation and expertise.

When talking with potential attorneys—and when you have found one who is compatible with your needs—always be sure to ask for an outline of expenses and if he or she is willing to notify you when the fees for a particular task will be exceeded.

Accounting

The accountant you select should, early on, be able to work with you on assembling your business plan, including your projected profit and loss statements, for financing.

Down the line, as your business is being established, the accountant will help you set up your books and, once in operation, should handle your tax returns, prepare financial statements, and offer financial advice regarding tax matters, cash flow, investments to maximize the use of profits, and the tax regulations regarding employees when you are ready to hire.

Fees

As with attorneys, there is a professional association in your state capital that certifies and maintains records on the reputation and fee structures of accountants.

The basis for fee structuring does vary slightly, however, with accountants. Some charge by the hour, others by the day, and still others work on a set monthly retainer based on the estimated amount of time they will be required to spend on your work.

Be prepared, however, to pay hourly fees that average between $25 and $100, depending on the accountant's expertise and location.

Insurance

Before setting out on your search for an insurance agent, it is advisable to have already established your business form and learned exactly what insurance the law in your area requires you

to carry (fire, liability, etc.) and, if hiring employees, what kind of program you'll be offering employees, as well as your own needs for medical and life insurance.

There are several types of insurance that you should carry for your coffee and tea store. *General business* coverage will protect against fire and theft, *general liability* covers accidents or injury to anyone while they are on your premises (at either home or retail location), and *product liability* covers against anyone getting hurt somehow with one of your products. If working out of your home, it may be possible to add these additional types of coverage to your existing homeowner's policy at a low annual cost.

You will also need to carry workers' compensation when you start hiring full- or part-time employees, unless they agree to work as independent contractors and take care of their own insurance and taxes.

The insurance agent you choose should be familiar with the needs of businesses and business owners, not just the standard life and disability policies. Your insurance needs will change as your business grows and expands (i.e., employee health, workers' compensation, etc.). At that point, you may want to consider key person coverage to ensure that your company can survive if a major partner or employee dies.

There are also a number of pension programs and stock-option programs available in the event you want to offer employees the incentive to increase their participation in the company in exchange for partial "ownership" down the line.

Fees

The fees for your agent's expertise are paid from your premiums, and there should be no extra charge for advice or administration of your insurance policies and programs.

Professional Services

Legal

Accounting

Insurance

11

TAXES, LICENSES, AND PERMITS

\mathbf{A}s a business owner, you are responsible for timely filing and payment of federal, state, and local taxes. Whether you have an accountant prepare your returns or you do it yourself, the task will be made much easier if you establish a systematic record-keeping system (as described in Chapter 15) and keep your records accurate and up to date.

This includes maintaining all written documents pertaining to the financial aspect of your business: invoices, bank statements, receipts of any and all business expenses, and deposit slips.

One of the easiest ways to keep control of the "paper dragon" is to set up a 9 x 12-inch manila envelope or a file folder for each of the following categories:

- ◆ *Paid Bills*—both personal and business

- ◆ *Sales Receipts* of every product you've sold or service job you've performed

- ◆ *Inventory* records based on ongoing inventory control and quarterly audits

- ◆ *Copies of Invoices* or billing statements that are paid, with a separate file for those still due you

- ◆ **Receipts** for miscellaneous cash purchases
- ◆ **Auto/Entertainment** receipts from travel and promotional activities

All of these documents must be kept for at least five years to substantiate deductions claimed on your income tax returns in the event of an Internal Revenue Service (IRS) audit. Make up new file folders or envelopes at the beginning of each year and store the old ones in a safe place.

Preparing income tax returns is not only a time-consuming task that can take you away from the important job of running your business, but also somewhat of an art form. Tax law is a constantly changing, complicated fact of life. It is strongly recommended that you have an accountant lined up to prepare your taxes and to keep you informed of any pertinent changes during the year.

Business Deductions

The deductions that you will most likely qualify for as a business owner include operating expenses such as telephone, postage, advertising, bank service charges, travel and expense of conventions, interest, dues to professional organizations, and subscriptions to magazines pertaining to your business, among others.

Operating your business from home allows you to deduct not only a substantial portion of your rent or mortgage payment, but also percentages of your phone and utility expenses.

If you have established your business at home, you will be able to deduct that portion of the house used exclusively for business, as well as a percentage of your costs for telephone service and utilities.

Again, because of the complexity and obscurity of many of the deductions, it is best to have a professional prepare your taxes

to ensure you get the full benefits to which you are entitled. Filing requirements will be determined by the type of business, the legal structure (sole proprietorship, partnership, or corporation), income from the business, your location, state and local laws, and whether or not you have employees.

For example, as the sole proprietor of your business you would probably only be required to file personal federal and state returns based on profit or loss, with the appropriate schedules for business expenses, pay scales, self-employment and estimated taxes, and local business license fees and sales taxes.

State Income Tax

Each state has corresponding filing requirements; however, form and schedule numbers vary. Contact your state franchise tax board or your accountant for details.

Local Taxes

Taxes will vary from city to city and county to county; however, you *may* be required to pay city income tax and local sales tax as well as real or personal property taxes. Check with your local government offices for specifics.

Licenses and Permits

To operate your business, you will need permits and licenses based on the requirements in your area and the type of business you are operating. You will probably be required to obtain the following documents regardless of where you live.

Local Business License

Basically, this is simply a fee paid to the city or county where you are located that allows you to operate your business in that area. Some cities will also require you to pay a percentage of your gross sales every year.

Fictitious Name Statement

This is a registration for protection of your business name. Filing the fictitious name statement will also involve a city- or county-wide search to make sure you are not duplicating an existing name. This can be done at your county hall of records. (See details in Chapter 12, "Naming Your Business.")

Seller's Permit or Resale Certificate

This is required only if you are going to be charging sales tax. Services are often exempt.

Health Permit

This is required only if you are preparing or distributing food in any manner, and involves an initial inspection and periodic follow-up inspections by health department officials.

Taxpayer Identification Number

This is available from the IRS in the case of partnerships, Subchapter S corporations, or corporations. Sole proprietors are required to have a taxpayer identification number if they pay wages to one or more employees or file pension or excise tax forms.

As you begin to outgrow your home, you will have to apply for licenses to operate a retail store. The kinds and types of licenses vary by state, so you will need to consult your local city or county offices for information. In most cities, one must apply for a resale license, which enables you to purchase products at wholesale prices and then resell them at retail prices. If you plan to include spirits in your coffee and tea store, you will need to apply for a liquor license, which can be costly.

Your local governmental offices or your attorney will be able to give you specific licenses and permits required, as well as the corresponding fees for each as they apply to your business.

Legal Structure

As a self-employed business owner, you are required to decide on a legal form of business for tax reporting purposes. There are four

basic classifications, as outlined. If, after reviewing them, you are still unsure of which way to go, it would be advisable to talk with a lawyer about the advantages and disadvantages of each structure for your particular business.

All businesses, regardless of size, are required to maintain detailed records and file the necessary tax returns. In a corporation, regular meetings must be held. The stockholders elect a board of directors, who establish and monitor corporate policy. The board selects corporate officers to conduct the day-to-day operations of the business.

Sole proprietorships are the most convenient and least complicated form of organization for new business owners, especially in the early stages. As your business grows, you will want to explore the options as a way of protecting your personal assets and increasing the potential for expansion capital.

Sole Proprietorship

This is the easiest to establish and is the structure many small business owners choose. A proprietorship is relatively free from government regulation, as the business has no existence apart from the owner. Profits from the operation of the business are treated as personal income for purposes of taxation, and your proprietary interest ends when you die or dissolve the business.

The major drawback of a proprietorship is that you are personally liable for any and all claims against the business and undertake the risks of the business to the extent of all assets, whether they are used in the business or personally owned. As a sole proprietor, you will be required to file self-employment tax returns and ordinarily would have to make estimated tax payments on a quarterly basis.

General Partnership

This is also an easy structure to set up and administer. Since responsibilities and capitalization are usually shared by two or more partners, taxation is based on each partner's share of business income and determined by their individual tax rates. Again,

claims against the business can be filed against personal assets, and financial liability is shared equally by all partners.

Limited Partnership

This structure can be established when one or more people are willing to invest cash or tangible property in the business with active participation in the daily operations. However, there must be at least one general partner who carries unlimited financial liability and who usually maintains a full-time managerial position within the company.

The limited partners are only liable for business debts up to the amount of their investment. Although a partnership is not a taxable entity, it must figure its profit or loss and file an annual tax return, which also becomes part of the partners' personal returns.

Corporation

In this structure, stock or shares in the business are sold to investors or stockholders, who then control the company. The advantage is that corporate stockholders are removed from any liability against personal assets. The most anyone can lose in the event of bankruptcy or a liability claim is his or her stock.

The privilege of reduced liability, however, creates paperwork (articles of incorporation and annual reports for the state tax commission and federal regulators), expenses (filing and licensing fees), and double taxation (the corporation is taxed on profits, while stockholders and elected officers are taxed individually on wages and/or dividend income).

Subchapter S Corporations

This legal structure has proved to be a real boon for small business owners who want the benefit of corporate protection from personal liability without double taxation. In a Subchapter S corporation, a maximum of thirty-five stockholders (who can be family members) report their share of corporate income on individual tax returns.

The corporation itself is generally exempt from federal income tax; however, it may be required to pay a tax on excess net passive investment income, capital gains, or built-in gains. To structure your company as a Subchapter S corporation, all of the shareholders must consent to the choice.

Notes

Key Points

Personal Thoughts

Additional Research

12

NAMING YOUR BUSINESS

As a pet owner, it is unlikely that you would give your German shepherd a name like Fifi. It wouldn't suit the dog's image, nor would it be appropriate. The same principle applies to choosing a name for your business.

The name you select for your business can be a tremendous asset when it defines the kind of image you want to project. You want the name to attract and appeal to potential customers, to be easily remembered over that of the competition's, and appropriate to the type of business you are starting.

Today's consumers are constantly bombarded with advertising as they go about their daily routines. Getting their attention, and holding it long enough for them to make an association between your business name and what it is you are offering, is imperative.

A memorable moniker can mean the difference between continued growth or a mediocre response from an audience victimized by information overload. (It is, of course, important to remember that your ultimate success depends on well-designed advertising, careful planning, and quality products and service.)

Sample Names

- ◆ A Cuppa Hot Joe
- ◆ Cafe Au Lait
- ◆ Coffee, Tea, and Thou
- ◆ The Coffee Bean
- ◆ The Coffee Grinder
- ◆ Espresso To Go
- ◆ Java Joe's
- ◆ More Mocha
- ◆ Pony Espresso
- ◆ Tea Time and Coffee, Too

If the name you select does not include your own name in the title, you must open a commercial bank account under the business name. Otherwise, checks made payable to your service could not be credited to your account, unless your bank is willing to let you sign a signature card on your personal account that covers the business as well.

If you are in a large metropolitan area, you might want to hire an attorney to conduct an investigation to check other uses of the name you select to ensure that you will not sued by any firm or large organization with the same name. If you found out that someone else has prior claim to the name after you had stationery printed, ordered a sign, and opened for business, you could lose a lot of money and would be forced to redo all of your promotional efforts.

The Simple Route

Many small business owners claim their image is enhanced by the use of a straightforward name that conveys their business concept but maintains a personal sense of identity—a factor that is especially important in today's high-tech world.

Historically, in developing business names, simplicity has scored the highest points. The name you choose should be short, to the point, and easy for customers to pronounce.

As an example, when Susan Hopper made the decision to open a preschool, she grappled with names like The ABC Center, Kiddie Land, and Now We Are Three. Finally, out of exasperation and because she felt these selections were either inappropriate or too cute, she decided to go with Susan's School, feeling it was easy to remember and conveyed exactly the personal, homey

atmosphere she wanted to create. She claims that 50 percent of the parents who picked her preschool for their children from the telephone book did so because it sounded friendly and sincere.

If, however, you do not want to name the business after yourself, you must be sure that what you do come up with is catchy enough to stand above the rest of the similar businesses in your area.

Brainstorming

Start by making a list of all the positive aspects of your business that you can think of, and call on friends and relatives to provide as many as they can come up with. Write down all the possibilities, no matter how funny or unusual they seem. A handy tool for business naming is the thesaurus, which will give you a vast number of options for commonplace names. Consider everything that springs forth from your imagination.

When you have created a list of likely candidates, get together with a group of supportive friends and family members and have a brainstorming session to either pick one of the choices you have come up with or to develop something from the ideas listed. Chances are that within a few hours, you will have a name for your business.

Catchy names are fun to design. However, make sure the name you choose isn't so offbeat, cute, or trendy as to risk slipping into obscurity as time passes.

The Fictitious Name Statement

You are required to file a fictitious name statement with the county clerk's office where you will be basing the business. While there, you should be able to do a countywide name check on the spot to see if there are any other businesses in the region using the name you have selected. The filing fee depends on where you live, and must be done within thirty days after you officially open your business.

It will also be necessary to publish the fictitious name in a local newspaper; the cost depends on the circulation of the paper.

The county clerk's office will advise you about specific requirements in your area.

If you are starting a business that will be operating in a broader market, statewide or nationally, it is important to have your attorney do a name clearance investigation, which can take from three days to three weeks.

Your Visual Image

After you have selected a name that reflects your business image, the next step is translating it into a visual symbol or logo (logotype) to serve as the signature piece for your business. Often this involves creating a visual interpretation of your company name, but in other cases a graphic symbol or trademark is designed to serve as identification.

Some established corporate trademarks are so familiar that you can immediately identify the company even without seeing or hearing its name.

A good example of this includes the logo of the dog with his head cocked to one side. The accompanying copy reads "His Master's Voice," and it's a good bet that you recognize this as the logo for RCA. Another effective logo is the avant-garde apple that identifies Apple computer products.

If you do not have the graphic skills necessary to design a logo with impact, get in touch with your nearest art association (listed in the phone book or available through local art supply shops or galleries) or call a nearby college or university. Ask the head of the art department if your design can be given as a class assignment or if he or she could recommend a student to do the job for a small fee. It will give students practical application, and the design can be used later in their portfolio. You can offer cash or a prize for the best design. The students will undoubtedly meet the challenge with enthusiasm and give you a number of good samples and ideas from which to choose.

Selecting a Typeface

Save sample logos and advertisements that use a typeface you like. Type is an extremely important element of logo design and can

also pinpoint the precise image you hope to express. Type not only presents the basic message, it can play a powerful role in the overall appearance of your logo and can actually create atmosphere. A chart of some of the more popular typefaces is included at the end of this chapter.

When deciding on a typeface for your logo, visit print shops or typesetting studios and look at their typeface books. They offer both the usual, functional varieties as well as a selection of unique typefaces that can really dress up your logo and subtly portray a specific personality such as dignified, fun, powerful, classic, or cutting edge.

Have the logo and your business information (address, phone number, etc.) set in more than the one typeface so that you can see how they will look when printed. Also ask to have them set in both small (10-point to 12-point for business cards) and larger (20- to 24-point for letterhead) versions. Once typeset, you will be able to make a final decision about which typeface suits the image you want to project.

Historically, in developing business names, simplicity has scored the highest points. The name you choose should be short, to the point, and easy for consumers to pronounce.

Typesetters generally have a minimum fee based on the amount of time they spend on a job, which can vary from $15 per hour in a small city to $50 per hour in a business area, as high as $100 per hour in major metropolitan areas. That's why it's important to shop around.

Word processing specialists or independent desktop publishers can also provide a variety of typefaces and formats at less expense. Since you will only be having a few words typeset, the time and cost required to set them in several different styles should certainly be affordable.

Business Cards and Stationery

The typeface and logo you eventually choose will be used on your letterhead; in your display and telephone advertising; on all promotional materials, including flyers, brochures, and announcements; on your sign; and on statements and invoices.

They will also be used on your business cards, one of the most inexpensive and convenient ways to inform people about your service or product. Once you have had cards printed, be generous. Give one to everyone you meet and always be sure to carry a supply wherever you go.

Most fast-print copy centers are prepared to help you if you decide not to design your own business cards and stationery. They have samples of business forms, letterheads, and cards with various styles to choose from. Make sure that your company name, logo, address, and phone number are included where necessary. If you have a fax and/or toll-free number, be sure those numbers are included. When someone looks at your card or letterhead, it must tell them instantly who you are, what your business is, and how you can be reached.

Sample Typefaces

Helvetica
ABCDEFGHIJ
abcdefghij

Helvetica Black
ABCDEFGHIJ
abcdefghij

Helvetica Condensed
ABCDEFGHIJ
abcdefghij

ITC Eras
ABCDEFGHIJ
abcdefghij

ITC Eras Demi
ABCDEFGHIJ
abcdefghij

ITC Eras Bold
ABCDEFGHIJ
abcdefghij

Korinna
ABCDEFGHIJ
abcdefghij

Korinna Bold
ABCDEFGHIJ
abcdefghij

Rose Round
ABCDEFGHIJ
abcdefghij

Lubalin Graph
ABCDEFGHIJ
abcdefghij

Britannic Light
ABCDEFGHIJ
abcdefghij

Britannic Bold
ABCDEFGHIJ
abcdefghij

Brush Script
ABCDEFGHIJ
abcdefghij

Optima
ABCDEFGHIJ
abcdefghij

Optima Italic
ABCDEFGHIJ
abcdefghij

Optima Bold
ABCDEFGHIJ
abcdefghij

Park Avenue
ABCDEFGHIJ
abcdefghij

Palatino
ABCDEFGHIJ
abcdefghij

Palatino Italic
ABCDEFGHIJ
abcdefghij

Palatino Bold
ABCDEFGHIJ
abcdefghij

Garamond Book
ABCDEFGHIJ
abcdefghij

Garamond Ultra
ABCDEFGHIJ
abcdefghij

Times Roman
ABCDEFGHIJ
abcdefghij

Times Italic
ABCDEFGHIJ
abcdefghij

ITC Fenice Regular
ABCDEFGHIJ
abcdefghij

ITC Fenice Bold
ABCDEFGHIJ
abcdefghij

ITC Fenice Ultra
ABCDEFGHIJ
abcdefghij

Eurostile
ABCDEFGHIJ
abcdefghij

Eurostile Bold
ABCDEFGHIJ
abcdefghij

Eurostile Extended
ABCDEFGHIJ
abcdefghij

Notes

Key Points

Personal Thoughts

Additional Research

13

DEVELOPING A BUSINESS PLAN

Developing your business plan is the most important process you will undertake in your career as an entrepreneur, regardless of the size or type of business you have decided to start.

A well-thought-out business plan will serve as a blueprint while your idea turns into a recognizable entity and as it grows into a stable and profitable venture. Too often we hear former small business owners say they probably could have made a success of their business if they had only known what to expect from the beginning...and that is where the business plan comes in.

Unfortunately, too many new entrepreneurs are unfamiliar with the importance of planning, or consider themselves an exception and feel they can succeed by winging it or dealing with problems as they arise. Not true!

Every business, whether a large commercial venture or a small home-based enterprise, needs to analyze its potential, examine its strengths and weaknesses, and determine the future of the organization. It works for the major corporations and it will work for you, especially once you become involved in the day-to-day operations of the business. Having a business plan will give you the freedom to follow the steps you have carefully laid out with

regard to budgeting, the success ratio of your product or service, the hiring of employees, and other growth decisions.

Advantages of the Business Plan

Once you have made the decision to become a business owner, you must devise a specific statement that clearly outlines *what* you plan to do, *when* you plan to do it, and *how* you will accomplish the short- and long-term goals.

Not only will this keep you on track, it will serve as an indicator to others of your sincerity and knowledge when you go out to find start-up or expansion capital, and it will serve as the foundation of your financing proposal.

The other advantage is that the actual task of putting your business plan together will help you define and clarify every step of your concept and, if done in a conscientious and objective manner, will point out potential trouble spots that can be addressed before they become major problems.

If all the necessary components are covered, it will put your business on the road to profit. It is a sure bet that, down the road, if you find your business is not generating the income you had originally projected, this is because you didn't include one or more of the basic business plan requirements.

Not a Guessing Game

Like any other major project, preparing a business plan involves time and research. It shouldn't be a guessing game. It will be necessary to ask yourself some very specific questions and to answer them thoughtfully and honestly. The business plan is your foundation, so build it carefully to ensure that it works at optimum efficiency for your needs. And make sure it is typed, orderly, and good looking so that you and others can recognize its importance in the professional scheme of your expanding operation.

An important aspect to remember is that your business plan is not cast in stone. In fact, what makes a business plan so great is that it invites revisions as your business changes and grows. This makes it a companion in your success and, by reviewing it regularly, a partner in your progress.

Working Tips

♦ Plan ahead. Allow yourself plenty of time to handle all orders without pushing yourself to exhaustion, especially during heavy gift-giving times of the year.

♦ Set up concise work schedules using daily and weekly planning calendars and create lists to make sure that you do everything that must be done, from filling special orders to contacting vendors for supplies.

The best way to approach your business plan is to take paper and pen and devote a few hours to coming up with some hard answers. Of course, you will want to condense your answers to fit into specific segments within the plan, including (in order of appearance) Concept and Feasibility, Legal Structure, Product or Service, Customer Base, Marketing and Production Goals, Personnel (your résumé and Entrepreneurial Profile and those of any other key personnel), and Financial Statements.

It is advisable to start each segment on a separate page and to create a table of contents to place in the front. Be sure it is well-written and organized, neatly typed, and bound in a report folder to preserve it and give it a professional look, especially when using it as a sales tool to convince lenders.

Key Questions to Ask Yourself

The first question you must ask yourself is: "Why am I interested in this particular business?" Probably the answer will be something to the effect of wanting to be your own boss and making money...independence and income.

This answer is fine as a personal goal, but it isn't going to be good enough if you are planning to approach potential lenders for funding. They will want to see an overview of your business concept, why you are convinced it will be successful, and where it fits in the scheme of similar businesses in your town or city.

The next question you must ask is: "What is my product or service?" This may seem like a ridiculous question since you know

your product or service, but it goes deeper. Your written response will include details about the service or a description of your product (preferably positive), with a focus on why customers will be inclined to work with you.

Additional questions to consider should include

- ◆ Why do I believe there is a need for my product or service?
- ◆ How do I plan to develop my business over the next five years?
- ◆ How much will I charge to ensure value to the customer and profit for myself?
- ◆ Who are my suppliers?
- ◆ Who are my customers?
- ◆ What equipment do I need to start the business?
- ◆ How much inventory and supplies do I need for start-up?
- ◆ What will it cost?
- ◆ Who is my competition and where is it located?
- ◆ What does the competition offer and how will I offer something different to attract customers?
- ◆ What changes are occurring in my marketing area that will impact my business in the future?
- ◆ What are my estimated sales figures for each of the next five years? (A guesstimate can be based on researching similar businesses in the area.)
- ◆ How will I advertise and promote my business and how will I pay for it?
- ◆ How and where is my product going to be manufactured?
- ◆ What is involved in the production—materials, labor, costs?
- ◆ Where will my service be performed?
- ◆ What equipment is required for my service and what are the costs for leasing versus purchasing?

- ◆ What are other overhead expenses (rent, employees, etc.)?
- ◆ How many people will be involved in the business and what are their qualifications?
- ◆ If I don't have employees, am I qualified to run the business myself?

By talking with people in similar businesses, as well as suppliers and direct competitors, and your local chamber of commerce, you will gather a great deal of information, both positive and negative, about your potential business. People love to talk about their successes and, if you ask in the right way, their failures as well.

If you operate like an investigative reporter for a few days while doing research for your business plan, it's guaranteed that you will obtain plenty of good, solid information. A Service Corps of Retired Executives (SCORE) representative through the Small Business Administration can also provide assistance or tell you about resources that can help you develop a realistic business plan.

Trade associations, listed in reference books available at your local library, can provide you with invaluable details on industry facts and figures, such as the percentage of gross sales that should be spent on advertising, the percentage that is typically paid for rent in your particular business, and how to price your product or service.

The final item to include in this section of your business plan, when and if presenting it for financing, is a personal résumé, designed to emphasize your general business management experience and your expertise within your specific area.

Describe the job duties for every job you have held, including any special aspects that pertain directly to the business.

If you cannot prepare the résumé, it is worth the minor fee to have it done professionally. The results will pay for the cost hundreds of times over.

Financial Statements

Once you have written your overview and description sheets, it's time to get down to numbers. This is the key to your business plan and, unfortunately, the area where many entrepreneurs get bogged down. But without an understanding about the numbers

*Show me a person with an obsession about
succeeding and a solid business plan, and
I'll show you a good risk.*

<div align="right">Anonymous Loan Officer</div>

involved, you can never expect to be a good manager and really shouldn't be surprised if you run into money problems within the first year.

Again, utilizing the resources indicated above—chamber of commerce, trade associations—you will need to work up your financial pages to include the following components, which most lenders will want to see projected for one and five years.

Projected operating expenses. This projected number includes materials, advertising, salaries for employees or outside labor, and other expenses directly related to the cost of doing business.

Estimate of gross (before tax) sales revenue. If the business is not yet operating, base your estimate on research figures from trade associations and what the local market dictates. If business is underway, base your estimate on how many items or hours of service you plan to sell and the average price you plan to charge.

How you arrived at the figures for these statements. Generally, you would base your figures on assumptions made about the number of months of operation, estimated number of sales, and the average amount versus the cost of each sale.

Cost of equipment and furnishings. Get estimated quotes, whether you are planning to purchase or lease these items.

Cost of materials. You need to know how much items will cost for production, if applicable, and for maintenance on equipment needed to run the business.

Additional operating expenses. You should research prices for rent, telephone and other utilities, business taxes and license fees, office supplies, and even decorating costs. Include a category called "other" to provide a cushion for unexpected expenses.

Balance sheet. This shows assets, such as equipment and operating capital you already have, and liabilities or debts and

expenses. If the business has not yet started, this would be a personal balance sheet indicating your net worth, listing all possessions of any value plus cash, stock, and other holdings, minus financial responsibilities.

Leasehold improvements. If you are planning to rent a commercial location or redesign a room within your home to be used strictly for business, estimate cleaning and restoration costs in this statement.

By investing time and energy into this portion of the business plan, you will absorb the numbers into your consciousness and be able to recognize, at a glance, when your costs exceed your profit margin or when you are in a position to start expanding.

If money matters are absolutely beyond your comprehension at this point, it would pay to hire someone to work with you to develop the financial pages of the plan. There are business consultants and accountants who will probably charge you a substantial amount, or you can approach the accounting or business department of the nearest college and see if there is a qualified student available to help you.

No matter who you find to assist you, however, be sure to stay involved in the process...the discipline and hard work will guarantee success.

Notes

Key Points

Personal Thoughts

Additional Research

14

FINANCING YOUR BUSINESS

Starting your business without having sufficient capital is setting yourself up for problems from the beginning. Undercapitalization is cited as one of the major reasons why businesses do not succeed; however, this simply boils down to bad planning.

If you research and record all the goals, marketing data, equipment and supply requirements, and financial needs of your venture before actually opening your doors, you will be able to see at a glance how much you need to get going, and *why you need it.* That way, there will be no surprises and no reason that your business should suffer from lack of capitalization.

It is important to have the financial resources to cover all your preliminary planning and start-up costs, including expenses incurred to research the feasibility of your business and expenses required to set up shop, from equipment and supplies to advertising and utility setup charges. In addition, you need a surplus to carry you over personally until the business becomes productive. The Cash Flow Statement, Start-up Costs Worksheet, and Projected Expense Statement forms provided in Chapter 5 will help you determine these expenses.

If, after developing your business plan (Chapter 13), you find that your personal resources are not enough to open the business, there are other options available.

The four most common methods include

♦ Starting the business on a part-time basis while holding a full-time job to cover expenses

♦ Taking on a limited partner

♦ Going to friends or family members for the money you need

♦ Applying for a loan through a commercial lender or the Small Business Administration (SBA)

There are, of course, pros and cons to each of these options.

Starting Out Part-time

Starting your business on a part-time or "moonlighting" basis is a decision that must be made based on the nature of the business. If you are planning to capitalize on your skills in upholstering, for example, you should have no trouble building up the business at night and on weekends.

It is perfectly feasible to start small, using your garage or a spare room as your production facility and purchasing an answering machine for potential customers to leave a message while you are at your regular job. When you get home, you simply call them back to discuss prices and arrange a time when it is convenient to pick up the piece of furniture to be upholstered.

On the other hand, if you are planning to start a temporary help agency, for example, it would be in your best interest to go into it on a full-time, dedicated basis, as your potential customers are going to want fast results and will call someone else if they are even slightly discouraged, such as getting a recorded message when they call.

Moonlighting will work with some businesses, but before exploring it as an option you must figure out if your limited availability will affect your credibility, if you really have the time and energy to work at a regular job and try to build a business (not to mention family responsibilities), and whether your ultimate goal is to be self-employed or just to earn a few extra dollars to supplement your base income.

Considering a Partner

A limited partner is one who will put up the money you need and step into the background to let you run the business the way you see fit. You must be sure, however, to have your attorney draw up a precise partnership agreement that covers every eventuality. Partnerships are typically entered into with the best intentions and the unwavering belief that the business will be successful.

Since this is not always the case—and even if it were—it is a businesslike move to ensure that such aspects as decision making, distribution of profits and losses, contributions of partners, and handling disputes and changes are outlined and approved by all the partners.

Approaching a Lender

Once you have your business plan and other paperwork prepared, decide which lending association you want to approach. Certainly, if you have a stable record with a checking or savings account at your regular bank or savings and loan, that is the place to try first. Set up an appointment with the bank manager or loan officer to make your request and explain why you feel your business venture is worth the bank's investment.

Be aware, however, that banks are more likely to provide you with a loan payable within five or ten years, as opposed to savings and loans, which are more interested in long-term loans, such as for mortgages. Credit unions operate in a similar manner to banks; however, you generally have to be a member. If you do belong to a credit union it could be your best bet, as it offers lower interest rates and can be more flexible in lending decisions.

If, for some reason, you do not want to run a loan through your bank, consider talking with other local small business owners. Very often, they can steer you to a regional, often independently owned bank or savings and loan that is empathetic to and supportive of new businesses. In that case, proceed as mentioned earlier and arrange a meeting with the manager or loan officer.

Present your case in a friendly yet professional manner. Be realistic and honest about your needs. Do not underbid because

you worry that you will not get a loan if you ask for too much. It is always better to start with a higher figure than you actually need so you have a strong negotiating edge.

In addition, most lenders have a pretty good idea about start-up and operating costs of new businesses and are much less likely to give you, and risk losing, a small loan for a business they know calls for more capital. They will be more willing to work with you if you are realistic and knowledgeable about your needs.

If, after your first try, the answer is no, ask for reasons why you are being turned down so you can restructure your presentation. Turn rejection into a learning tool to redefine and polish your material and to develop new negotiating strategies. There are always other lenders you can approach, and the law of averages dictates that you will get your loan if the idea is solid and it is apparent that you have researched the feasibility of starting a business in your particular area.

Money brings some happiness. But, after a certain point, it just brings more money.

Neil Simon

Financing Methods

"Friendly Financing"

The third option, raising capital through friends or family members, is probably one of the methods most often used. The advantage of getting a loan from a personal contact is that he or she knows you, undoubtedly trusts your ability to make the business go, and won't require much in the way of substantiating paperwork, such as complex loan applications, financial statements, etc. In addition, you will most likely be able to negotiate a very low interest rate on the loan.

The major disadvantage, according to entrepreneurs who have taken this route, is if the friendly lender decides that he or she wants to provide input on the care and maintenance of your

business. This problem, however, can be eliminated by a thorough business discussion prior to accepting the loan. In other words, choose your investor carefully!

The second problem has to do with repayment of the loan. Even though you have a loose agreement in writing with your lender, because of friendship or family ties there may come a point when Uncle Bill needs that $10,000 *tomorrow* to take care of a personal obligation. You can't possibly come up with the money overnight, Uncle Bill gets angry, and that whole side of the family turns against you.

The flip side of the coin is if the business fails and you are unable to pay Uncle Bill the $10,000 he put up. There are any number of unpleasant situations, so you must be sure in the beginning to think about the importance of the relationship you have with the potential lender, how the best and the worst of circumstances would affect your relationship, and whether you then could justify asking for money.

Commercial Lenders

If you are not able to, or if you decide against approaching friends or relatives for financial assistance, the next step is a bank, a savings and loan, or a credit union. Before approaching any of these commercial lenders, you must have carefully developed your business plan, which will include the following documents.

1. A résumé or statement outlining your background and capability to operate the business, plus a similar statement about any key employees or partners in the business.

2. A statement of business and personal goals.

3. A description of the business, including market research for your product or service.

4. Details on how the business is going to be structured (sole proprietorship, partnership, corporation).

5. A projection of profit and loss for a minimum of one year (five years is preferable), which means you will have investigated how similar businesses are doing.

6. An outline of how much money you need–and why–to keep the business solvent and to support yourself and your family for at least a year.

In addition, you will be required to provide a personal balance sheet that lists your assets (real estate, a car, valuable personal property) and liabilities (mortgage payments, credit card debts) along with a credit application that outlines your personal financial history, so a determination can be made on your ability to pay back the loan. The lender will follow through by requesting a credit report from an independent agency to help with the decision.

The main point to remember when applying for a loan with a commercial institution is that lenders aren't as concerned about the amount loaned as they are about how and when they are going to get the money back!

Paying Back the Loan

When you apply for financing, the burden of proof as it relates to repayment rests with you. No one would knowingly grant a loan to an individual or a business that he or she had doubts about. As a borrower, your responsibility is to show that you will be able to pay back the loan according to the terms agreed to by you and the lender. If you have a good credit history, a credit report will indicate as much.

The Small Business Administration

The Small Business Administration (SBA) often goes where no other lender will tread and, as such, is a lender of last resort. The SBA is a government agency that is well known for providing financing to entrepreneurs who have been repeatedly turned down by commercial lenders, which, in fact, you generally must be before the SBA will consider backing you.

After your loan request with a commercial institution has been denied, you can file an application with the nearest branch of the SBA. It is a good idea to make an appointment with a Service Corps of Retired Executives (SCORE) representative (a man or woman who volunteers his or her time to the SBA-SCORE program to advise new and established business owners). Your SCORE

representative will lead you through the complex paperwork required by the SBA before it makes a decision.

In addition, SCORE volunteers are usually straightforward, knowledgeable men and women who will walk through your business plans with you and offer constructive suggestions. Once the paperwork is completed, a commercial lender will make the loan under the SBA Lender Certification Program, knowing that the government is willing to insure it.

We recommend this option only after you have been turned down by three or more banks, because of the time factor involved in gaining approval and also because of the extensive follow-up reports required by the SBA. It is, however, a viable option and one that has helped thousands of dedicated entrepreneurs realize their goals.

Venture Capitalists

Money is available to businesses that are already established and are seeking working or expansion capital from groups of investors known as venture capitalists. These groups can vary from a few local businesspeople with money to invest to major investment companies connected with large corporations or financial organizations.

Venture capital is not like a straightforward business loan. It is usually dependent on a minimum $100,000 investment and, therefore, is not suited to every business situation. Typically, venture capitalists are interested in companies that have a track record, a proven position in the market, and a solid growth projection.

But, like a bank or other lending association, venture capitalists want to see a written business plan and a prospectus of future projections. They are looking at your background, the market, the kind of funding you want, and your past financial record.

Since venture capitalists are looking to earn from 10 to 15 percent on their investment over a relatively short period, they will want to spend a great deal of time talking with you and your associates, customers, and suppliers.

Before considering venture capital, we advise discussing it carefully with an attorney who can help you investigate different

groups to figure out the best investment structure, and who can work with the group's attorney on drawing up an agreement that protects you. Many venture capitalists will want to own part of your company.

This is an option to be considered only when your company is well established and undergoing rapid growth pains and should be approached with great understanding of the situation.

Other Financing Options

Loan Companies

Loan companies are good sources of funding. However, interest rates are high, and they will generally want to have substantial collateral, such as the equity in your house, on record before making a loan.

Insurance Companies

Your insurance carrier may be willing to make an investment in your small business, using your insurance policy as collateral, or you may even have enough cash value in your policy, depending on the face amount, to provide substantial start-up capital. If this is the case, you will only be required to pay quarterly or semiannual interest payments on the cash value you have taken out.

Factoring

In this instance, a factoring company "buys" your accounts receivable and advances you a percentage of the full amount due. This is a viable option for well-established service companies that work on a billing basis.

Cosigner

If you have a relative or friend who is already an established business owner or, at least, a homeowner with a solid credit rating, it might be worth your while to ask if he or she would cosign on a loan application with you. Although you are still responsible for repayment of the loan, the bank is assured that, in the event you default for any reason, the cosigner will guarantee the obligation.

It is sometimes difficult to find someone who will do this, but again it doesn't hurt to ask, especially if it is a last resort.

Starting Small

Even if you know your particular business is valid and that you have the ability to make it succeed, be certain that your business plan is realistic. If you have chosen to start a business on a grand scale but have minimal capital and little business experience, it may be best to begin a smaller, less elaborate operation at first.

You'll require less "seed money" and will be in a low-risk position while learning the operations of a business and seeing if you can handle all the variables of business ownership while making it grow.

Smaller businesses have proven to be a great way to learn successful methods, as well as a vehicle for ironing out the many small, often overlooked, details. The profits you gain from a smaller venture can be used to expand or invest in bigger business ideas. And an added bonus is that when you are ready to approach investors or lending institutions, you will be able to show them that you already have a solid track record and a working knowledge of business procedures.

The journey of a thousand miles begins with a single step.

Chinese Proverb

What to Do When Asking for Money

1. ***Be sure to ask.*** This may seem like a gross statement of the obvious, but you would be amazed at the number of small business owners we've heard about who never ask because they are afraid of being turned down.

 Unless you are independently wealthy and pursuing your business as a humanitarian effort, it is unlikely that you are in a position to run your business and earn enough money to support you, your family, and the

operation—especially during the first year. Remember the adage: It *takes* money to *make* money.

If you run a low-budget business, you will probably get a low-budget response. If you are determined to make it work, be sure you have sufficient capital to make it work the right way.

Fear is often a factor: "I don't want to ask, in case I'm told no." Well, that's the worst thing that can happen. But if you persevere and are serious about your venture, someone will inevitably say yes!

And don't overlook friends and family; they can be your most ardent supporters if you have given them reason in the past to believe you are responsible and determined to succeed.

2. ***Know how much you need.*** Lenders are familiar with the financial demands of business operation and will respect your request if you have obviously done your homework and can talk intelligently about your needs.

3. ***Be direct and confident.*** If you believe in your business and in your ability to make it work, others will be convinced as well. Never apologize for mistakes you feel you have made in the past.

 Simply present the facts honestly, even if it includes revealing an error in judgment you have made somewhere along the line, and assure the lender that he or she will be making a smart decision by investing in you.

4. ***Think positively.*** If you need $50,000, ask for $50,000. Never underestimate the potential to provide. Remember: It is easier to negotiate and deal with one lender for a single amount than it is to keep paperwork and relationships strong with several, all of whom have contributed to the pot.

5. ***Ask again.*** If they trusted you once and you have lived up to the stipulations of the contract, ask again—that goes for commercial lending institutions as well as friends and relatives. When it comes to borrowing money, a proven record is what it's all about, and if you have established

yours, keep it active. It will be your passport to profit when the chips are down.

6. **Know when to borrow.** If you have worked out your business plan and know you can survive while getting the business off the ground, start exploring your financing options ahead of time. Don't wait until the last minute when you will be forced to act and will have to take a less-than-favorable opportunity. The same theory applies if your business is already established. By examining your financial position on a regular basis, you will be able to project how much you will need at a given point for expansion purposes. Always be aware.

7. **Don't borrow if it is not necessary.** Many businesses can be started for under $1,000. This is called "starting on a shoestring" and can be done with a variety of businesses. Services, for example, often rely strictly on the owner's knowledge and expertise and can be set up quickly and inexpensively.

If this is the case with the business you have in mind, then try to avoid borrowing money, which can be an expensive and timely proposition.

Establishing Credit

Is it possible to get a loan even if you have never established credit? Yes, it is. If you prefer to pay cash rather than incur high interest charges on loans or credit cards, you can still qualify for a loan based on personal assets or by having a friend or relative with a good credit rating cosign for you on the loan. This puts the obligation on the cosigner, so be sure the terms of the loan are clearly spelled out in a written agreement to the satisfaction of everyone involved.

But, if your personal assets are minimal and you cannot find a cosigner, your best bet is to put off starting the business for about four to six months while you establish credit. The best place to start is with a major department store. A store will issue you a credit card based on a very simple examination of your income and employment history. When you receive the card, charge about

$100 worth of merchandise and pay it off according to the schedule provided. Within a few months, you will have proven yourself to be credit-worthy, which will greatly improve your chances of getting a loan from a lending institution.

Another way to establish credit—and credibility—is to open a checking account at the bank you have decided to approach for a loan. Banks generally require a minimum deposit of between $50 and $100. Make it a point to meet the branch manager and/or the loan officer and establish an ongoing relationship with them by stopping by to say hello when you are in the bank.

Within a few months, apply for a small personal loan, working with your new acquaintance, of course. Make your payments according to the prearranged schedule. Then when you are ready to request a more substantial amount of money to cover your start-up expenses, you will be recognized as a customer with a loan history at that institution.

Notes

Key Points

Personal Thoughts

Additional Research

15

RECORD KEEPING: YOUR BUSINESS LIFELINE

The motivating factor in any business is profit, which can be explained as the money left over after all the bills–everything from supplies to rent to salaries to taxes– are paid.

Building a profitable business is not something that can be left to chance; it must be planned, and a systematic method of record keeping must be developed to help you monitor income and expenses.

You should expect that during the early days of your business, your profits are going to be minimal as you become established. But it is possible, with even simple record-keeping procedures, to prepare yourself for lean periods and control day-to-day expenses to ensure that you are, at least, breaking even. In addition, financial records are required for tax purposes, and dealing with them systematically can help you avoid an incredibly overwhelming task at tax time.

Record Keeping Can Be Simple

Some people cringe at the thought of record keeping or feel it is a waste of valuable time. Usually, these attitudes are based on a lack

of knowledge and the feeling that it is a difficult job. There is, however, no other way to analyze your cash flow and make sure you are pricing products or services high enough to realize a profit.

In truth, record keeping is not a complicated process. If you have ever balanced a checkbook or planned a household budget, you have performed several of the same steps that are necessary for implementing a bookkeeping system for your business. And the good news is that keeping business records does not have to be either complicated or time consuming.

We know of entrepreneurs who opt for total simplicity by using the "shoebox" method—every sales record, receipt for expenses, and bank statement gets tossed into a box. This system has two distinct drawbacks. One may not become apparent until tax time, when you attempt to wade through and organize each sheet of paper to prepare your tax return. The other, more critical drawback is that it is virtually impossible to maintain an accurate picture of your financial situation when you stockpile, rather than record, business transactions.

In order to understand your cash flow, it is important to be able to see monies earned, monies paid out, current balances, and outstanding debts.

In fact, you should be able to answer the following questions with just a quick review of your records.

- ◆ What was my income last year (week, month)?

- ◆ What were my expenses?

- ◆ How do income and expenses compare with last year (week, month)?

- ◆ What was my profit (loss) last year (week, month)?

- ◆ Where can I cut back on expenses?

- ◆ Who and how much do I currently owe on outstanding debts?

- ◆ Who owes me money and how much?

- ◆ What are my assets, liabilities, and net worth?

- ◆ Is my inventory in line with demand?

♦ How much cash do I have available? How much credit?

♦ Am I able to pay myself this month (week)?

♦ Are my figures in line with projected financial goals?

The records you need to be able to answer most of these questions are your cash journal, a balance sheet, and a bank reconciliation. A simple single-entry system, as indicated on the following pages, where you can record disbursements (cash paid out) and receipts (cash taken in) will form the base of your record-keeping system.

> *If you can organize your kitchen, you can organize your life.*
>
> Dr. Louis Parrish

Double-Entry Bookkeeping

You may want to consider a double-entry system, which involves recording each transaction twice; once as a debit (simply the left column of the ledger) and once as a credit (the right column of the ledger). For example, if you were to sell a product for $100, the transactions recorded in a double-entry system would be as follows: The $100 would be written as a *credit* in your sales account, since merchandise is going out of the business, and $100 would be recorded as a *debit* in your cash account since money was coming into the business.

This is a complex and time-consuming process that is often best left to an accountant, as he or she will need the information to create a monthly trial balance and other financial statements, including your year-end tax reports.

Single-Entry Bookkeeping

You can, however, set up a simple single-entry system, which will tie in directly with your accountant's requirements, or check out the standardized bookkeeping systems available at office supply or stationery stores, which provide all the necessary forms and documents in a bound book.

Bookkeeping Debits and Credits

Debits

◆ Cash receipts

◆ Purchases

◆ Expenses, such as rent and wages

Credits

◆ Cash payments

◆ Sales of services or merchandise

◆ Earnings, including interest

A system should contain forms for recording monthly income and expenses, summary sheets from which you can create a balance sheet, and listings of legal deductions for income tax reporting. In addition, the trade association for your field should be able to provide you with systems developed exclusively for use in the industry, which you can use or adapt according to the specific circumstances of your business.

The final method is to purchase a cash journal book and set up your own monthly system. (We've used Office Assistance, a small secretarial service that has been in operation for one month, as an example.) Any of the above-mentioned methods are acceptable, as long as you understand the entry process and can read the results.

Make Record Keeping a Daily Task

Short of paying someone else, the easiest way to be sure your records are kept up to date is to incorporate the task into your daily or weekly routine. Many small business owners make it a habit to enter their sales, expenses, and other financial information at the end of each working day. It keeps them continually aware of their financial situation and ensures that there will never be any cash-flow surprises.

The process shouldn't take more than fifteen minutes for normal transactions, and it will save hours of pencil pushing and frustration down the line. What's important is that you'll always know where you stand financially.

Setting Up the Books

Now, using Office Assistance, we will examine the various elements required for basic record-keeping duties.

Office Assistance has been in business for one month. Two months ago, Bill Miller, president of Office Assistance, opened a new business bank account with $10,000.00, his start-up capital from a personal savings account.

At the same time, he rented a small office in a downtown building for $350.00 a month, but had to pay the first month's rent, and a deposit of the last month's rent, for a total outlay of $700.00.

His fictitious name statement, which he got approved through the local county clerk's office, cost $10.00, and publishing it in a regional newspaper cost $45.00.

The initial month's lease and a deposit on a computer cost him $275.00, plus $50.00 for a maintenance agreement. However, he will own the $2,000.00 computer when his payment schedule is completed.

He found a brand-new calculator at a garage sale for $25.00 and is going to use a desk, table, lamp, and chairs brought from home (value $350.00) to decorate the office. Phone-line installation was $150.00, and he purchased a two-line telephone for $79.50.

An artist friend designed his logo and letterhead on a computer for only $25.00 and a $6.95 lunch. He had his stationery ($35.00), business cards ($60.00), and brochures ($23.50) produced through a local copy shop for a total of $118.50.

A 2 x 2-inch display ad in the local newspaper will cost him $370.00 for a week, and he is planning to mail 100 of his brochures to local businesses selected from the phone book. Stamps will cost $29.00 for the mailing. Office supplies, including typing paper, staples, paper clips, etc., set him back $45.00. A journal for record

keeping cost $7.95 and a packet of invoices $5.95, for a grand total of $58.90.

Since he doesn't know how to type, Bill hired his niece to work 20 hours a week for $10.00 per hour on an independent contractor basis so he doesn't have to pay social security or unemployment taxes.

During the first month, he has billed and been paid $700.00. However, he has two clients who still owe him a total of $400.00. Bill dutifully records information in his cash journal at the end of each working day. He uses source documents, including his checkbook register, receipts from cash purchases, and billing invoices as the basis for his entries. The following cash journal illustrations are for May (prior to opening the doors of his business) and June (his first actual month in business).

Bill's expenses for May and June were $3,108.75. Of course, part of that is for start-up expenses, such as deposits on his rent and computer, installation costs, and one-time fictitious name filing and publishing. His income for the first month was $700.00.

By deducting his expenses from his income, he can see that, at the moment, his business is showing a loss of $2,408.75.

Office Assistance
Cash Journal for May

Date	Check or Invoice #	Detail	(Debit) Expense	(Credit) Income
5/1	100	Rawlins Real Estate (rent & dep)	$700.00	
5/5	101	County Clerk (fictitious name)	10.00	
5/7	102	The Herald (publishing FNS)	45.00	
5/9	103	Ed's Keyboards (IBM l-mo. & dep)	275.00	
5/9	104	Ed's Keyboards (maint. agreement)	50.00	
5/12	105	Mary Smith (calculator purchase)	25.00	
5/18	106	Telephone Co. (line installation)	150.00	
5/20	107	Phone Store (2-line phone)	79.50	
5/22	108	Ray Brown (logo design)	25.00	
5/24	Cash	The Hungry Dog (lunch/Ray Brown)	6.95	
5/28	109	The Copy Spot (brochures, cards)	118.50	
5/29	110	The Herald (advertising)	370.00	
		Total Expense & Income (May)	**$1,854.95**	**$00.00**

Office Assistance
Cash Journal for June

Date	Check or Invoice #	Detail	(Debit) Expense	(Credit) Income
6/4	111	U.S.P.O. (stamps for mailing)	$29.00	
6/6	112	Office Stationers (supplies, invoices, etc.)	58.90	
6/7	A1	W. Smith		$62.50
6/8	A2	Art Association		112.50
6/9	A3	T. Williams		22.00
6/9	113	Judy Miller (typing fee)	200.00	
6/10	A4	Bank of Cutterville		75.00
6/10	A5	WKTR-FM		120.50
6/13	A6	J. Johnson		43.50
6/15	114	Rawlins Real Estate (rent)	350.00	
6/15	A7	C. Lewis		73.50
6/15	A8	R. Swell		90.00
6/16	115	Judy Miller (typing fee)	200.00	
6/19	A9	W. Smith		52.50
6/23	116	Judy Miller (typing fee)	200.00	
6/26	117	Phone company (bill)	15.90	
6/27	A10	K. Black		48.00
6/30	118	Judy Miller (typing fee)	200.00	
		Total Expense & Income (June)	**$1,253.80**	**$700.00**

Although Bill has only been in business for a month, he is curious about his company's financial worth and decides to work up a balance sheet to get the answer. The calculation, as indicated in the example, is the amount owned (assets) minus the amount due to creditors (liabilities), which equals his worth.

Balance Sheet (as of June 30)

Assets		Liabilities	
Cash on hand & in bank *(Capital balance & June Income)*	$6,191.25	Ed's Keyboards *(Balance on IBM)*	$1,725.00
		Unpaid rent *(July)*	350.00
Office equipment *(Includes full value of IBM even though not paid off)*	2,000.00	Taxes *(estimated)*	75.00
Office furniture	350.00		
Accounts receivable *(outstanding invoices for work already done)*	400.00	**Liabilities**	$2,150.00
Total Assets	**$8,941.25**	**CAPITAL**	$6,791.25
		Total Liabilities	**$8,941.25**

The figure Bill is most interested in is the CAPITAL amount in the Liabilities column. This is the amount remaining after what Bill owes is subtracted from his current assets and is what his business is worth at the end of June. In other words, if he decided to try to sell his business right now, he could realistically ask that amount as a selling price. Of course, Bill probably wouldn't get that amount because he has not yet become established enough to warrant someone buying the business, unless he or she was looking for a "turnkey" operation—in other words, a business to just walk into and get going immediately.

This information is valuable when Bill goes to apply for expansion capital or for credit on future purchases he plans to make. His balance sheet will change each time he prepares it (probably quarterly in the future) as business improves, bringing in greater income and reducing his debts.

In the meantime, the balance sheet gives Bill a tool to use when comparing the financial standing of his business this month against future months and years. It also keeps him current on what he owns, to whom he owes money, and his major sources of income.

The same procedure is used in developing a personal balance sheet, which possibly would be needed to establish credibility when applying for a loan. Assets would include furniture, automobiles, jewelry, a home, and other tangibles, while liabilities would consist of outstanding loans and other major debts.

Bank Statement Reconciliation

Another important step that Bill must handle monthly is reconciling his bank statement against his checkbook register. He simply marks off the checks in his register that have cleared per the bank statement, notes the deposits that have been credited, and deducts any service charges for the previous month from his balance.

Bill then adds up all the outstanding checks—those listed in his register that have not cleared by the closing date indicated on the bank statement—and *deducts* them from the balance indicated on his bank statement. He adds up any deposits that have not yet been credited to his account and then *adds* them to the balance, as follows.

Balance per bank statement	$5,982.15
Plus: Deposits not credited	+325.00
Minus: Outstanding checks	-115.90
New Balance	$ 6,191.25

The new balance figure should match that listed in his checkbook register and, in this case, it does. If, however, the statement and the register did not reconcile, Bill would have a customer representative at the bank review his statement and banking activity for the past month.

Coffee and Tea Store Record Keeping

A common mistake in the food service industry is a lack of understanding about the basic records that must be kept by owners. They are important factors, ranking next to good food, service, etc. Maintaining accurate figures from the start of your business is one of the most important steps to success.

There are a variety of forms to help you keep records of your daily business transactions. These are available at any complete business supply store. You will want a register closeout sheet for when you open and close your register every day. Some good register procedures are as follows:

1. If money pickups are made at busy times to remove excess bills from the register, a slip should be left in the register stating who made the money pickup, the amount, and to where it is being transferred.

2. Leave the register drawer open at closing time. In case of a break-in, the register won't be destroyed by someone trying to break it open.

3. A count of how many customers you serve a day can be a valuable tool for you. This information can be taken from your register, daily or hourly.

4. You must make a bank deposit every day. Do not make it at the same time every day, to avoid being robbed. Use a bank deposit form for this transaction.

Notes

Key Points

Personal Thoughts

Additional Research

16

PRICING FOR PROFITS

One of the toughest problems facing small business owners is establishing prices that the market will bear and that will cover overhead and guarantee a profit.

Often new business owners give the business away to get sales, but this is not an advisable practice. Realistic pricing indicates your confidence in what you are selling, and if you value your service, so will the customer.

Today's consumer realizes that no one gets something worthwhile for nothing, so don't be afraid to establish prices that will work toward your profit goals.

Guidelines

Several factors must be taken into consideration when setting prices:

1. *The cost of goods sold.* In the case of a retail or wholesale operation, this is the amount originally paid for the goods. For a manufacturer, it is the cost of producing the goods.

2. ***The nature of the product or service.*** Uniqueness and demand come into play here. In the case of goods with a stable level of demand, such as bread or auto repair, raising or lowering prices will have little effect. However, when demand is high for goods that are hard to get, the price can realistically be set anywhere the business owner wants. Such is often the case with restaurants featuring an exclusive or unusual menu.

3. ***The competition.*** It is important to recognize what your competition is charging, for often this will guide pricing within a certain region. However, if a competitor is charging what you feel is an unrealistic price—either more or less—for a product or service, you owe it to yourself to find out why. Then set your prices according to all of the factors outlined here. Even if your prices are higher than the competitor, consumers will pay the cost if you offer something special, such as a friendly atmosphere, convenient hours, follow-up service or maintenance, or some other benefit not provided by the competition.

4. ***Your company policy.*** This encompasses a number of things, including your location, your position in the marketplace, and additional services offered. It also takes into account your personal philosophy about business and your role in it.

5. ***Market strategy.*** Should you go for large volume at low prices or low volume at high prices? That is the bottom line in considering market strategy. As a small business owner, you will likely opt for low volume and higher prices, since the alternative involves having the resources, including labor, display room, distribution channels, etc., to move large volumes of product or perform major service tasks.

6. ***Customers.*** What will the market bear? In other words, what are your customers willing to pay for your products or services?

People expect prices that are fair; if you are planning to charge overinflated rates you had better be a topnotch salesperson or have something so incredible and unusual to offer that the price won't matter.

Although there are differences between establishing prices for retail operations, wholesale products, manufacturing, and services, the basic formula for price setting is

Labor + Materials + Overhead + Profit Margin = Price

However, before setting prices on goods or services, it is extremely important to understand the concept of the break-even point. Many small business owners operate on an overall profit-loss basis without realizing the importance of cost accounting. Being aware of such factors as your break-even point, markup, and profit margin can tell you which areas of your business are profitable and which are causing a drain.

Understanding Break-even

The break-even point is the minimum amount you must charge in order to cover all expenses incurred for the production and promotion of your goods and/or services *without losing or making money.* In other words, any income that is above the break-even point is considered to be profit and anything below it is a loss.

To find your break-even point, you must first total all of your operating costs, including materials and labor, equipment lease or purchase payments, advertising, utilities, office supplies, and any incidentals such as gasoline, maintenance, and postage.

Generally, this is computed for a particular period of time, such as six months or a year. However, if your business is still in the early stages of operation, you can use the estimated figures on your projected expenses statement and guesstimate costs for materials and labor.

For example, the monthly expenses for your hypothetical cake-decorating business total $300.00. You want to know what

the break-even point would be if you sold an average of 20 cakes per month. The calculation is as follows:

$$\frac{\$300.00\ \text{(expenses)}}{20\ \text{cakes}} = \$15.00$$

In order to break even, that is, to operate without losing money or realizing a profit, you must charge a minimum of $15.00 for each cake sold.

The same process can be used to analyze the break-even point on a weekly basis. First you determine your annual expenses by multiplying the $300.00 by 12 months, which would give you $3,600.00 per year.

The calculation to find the weekly break-even point is:

$$\frac{\$3,600.00\ \text{(expenses)}}{52\ \text{weeks}} = \$69.23$$

Therefore, you must earn $69.23 per week to operate the business without losing money and without realizing a profit.

Stay Informed

Knowing your break-even point is one of the greatest favors you can do for yourself as a business owner. It tells you how much you must charge for your products or services and is invaluable in setting prices to realize a profit.

Keep in mind, however, that the break-even point is a variable figure. Since it depends on production and overhead costs, plan to reevaluate periodically to make sure your prices reflect any changes.

Labor Costs

Labor costs, obviously, are the expenses incurred for the actual work done to manufacture or sell a product or to perform a service. Think of them as wages or salaries.

Small business owners often end up working for free because they fail to set a wage for themselves. Despite the fact that you will want to reinvest all of the income received back into the business for a while, it is imperative that you establish a fixed salary amount for yourself when figuring operating expenses.

If you have set aside a survival fund to carry you through the first six months of operation, you may want to defer your salary until the business becomes solvent; however, you should still figure the amount into your expenses. Otherwise, you may find the prices you set are too low to justify making a profit from the onset.

It is much easier to set realistic prices from the beginning than it is to raise them later in an attempt to make up the difference. Remember, your time and skills are the cornerstone of your business, so think of paying yourself as you would any valuable employee.

To do this, you must know what your costs are, or at least have a solid idea of projected costs if you are still in the process of planning your venture.

By not setting prices that are too high or too low for the product or service you are selling, you will be assured of a favorable position in the market and a healthy share of the wealth.

Setting Retail Prices

If you are manufacturing items to sell at retail prices, without using a middleman, use the following formula as a starting point when establishing a selling price for your goods.

$$\frac{1}{3} \text{ Labor} + \frac{1}{3} \text{ Materials \& Overhead} + \frac{1}{3} \text{ Profit} = \text{Selling Price}$$

You need a foundation. One place to start is with your labor costs. If, for example, the monetary value of your time and effort (labor) in producing an item were $6, you would allocate $6 for material and overhead and an additional $6 as profit for a total selling price of $18.

If materials and overhead are costing more, you can (a) boost the price accordingly, (b) review your expenses and find ways to cut back on material costs, such as finding a less expensive supplier, or (c) utilize a portion of the profit margin to cover the balance. Of the three options, (b) is the best way to go.

You must also remember that when you are producing mass quantities of an item, your costs will be reduced because of price breaks on supplies and reduced labor costs per unit. You're then able to structure your selling price according to previously mentioned pricing factors, such as competition and demand for the product.

If you are starting a retail business that relies on selling products you purchase wholesale, you may have some of your prices set for you by what the competition is doing, or by recommendations made by the wholesaler. However, there will still be items you must price yourself.

This will involve understanding the principle of markup or gross margin—the difference between the cost of goods sold and the selling price, taking into consideration sale markdowns, shortages, and discounts to employees. Generally, markup is stated as a percentage of retail price.

Money is the sixth sense which makes it possible for us to enjoy the other five senses.

Richard Ney

For example, if a manufacturer sells dresses to you at a wholesale price of $12.50 each and you sell them for $25.00, you would have a 100 percent markup (a 50 percent gross margin).

If you know the cost of goods and the average amount of markup you need to operate profitably, it is relatively simple to determine a price by using the following formula:

$$\frac{\text{Cost of goods}}{(100 - \text{markup } \% \times 100)} = \text{Retail Price}$$

For example, if you purchased a gross of rubber ducks for a total price of $172.80 and had determined that you needed a gross margin of 36 percent to operate profitably, you would calculate the expected profit as follows:

$$\frac{\$172.80}{(100 - 36 \times 100)} = \$270.00$$

When you divide the retail price ($270.00) by the number of items in a gross (12 dozen or 144 things) you get a unit price of $1.88, which you would probably raise to $1.98, depending on the market, to improve the profit margin slightly and provide leeway for markdowns during sales, etc.

Specialty items, such as antiques, artwork, imported goods, and handcrafted items can be priced higher according to current value and what the market in your area will bear. Their markup usually ranges between 200 and 300 percent.

Setting Wholesale Prices

Operating as a wholesale manufacturer greatly reduces your selling and administrative costs, because you are passing your products on to someone else to sell. Your price to retailers should thus be approximately 50 percent less than the suggested retail price.

In actuality, you are providing retailers with a discount because of their willingness and ability to promote your product.

The formula for wholesale pricing includes your profit margin + labor costs + expenses, which will be appreciably lower than for a retail operation because of savings on advertising, display equipment, and maintenance, but which must include warehousing and marketing.

Pricing is typically more competitive at the wholesale level than at any other and is almost always the determining factor in whether your products are purchased or not.

Also affecting wholesalers are other wholesalers offering competitive pricing, middlemen buying large quantities at low prices, and supply and demand factors.

As a wholesaler, you are in a position to vary prices according to the size of orders and your ability to negotiate with your buyers. However, it is a good idea to develop a solid markup base from which to operate your business. This will allow you the flexibility to offer maximum and minimum prices for each item based on quantity buys.

Pricing is a crucial aspect of managing your business. Since you are in business to make a profit, it is important that you set prices that will result in the greatest income.

To do this, you must know what your costs are, or at least have a solid idea of projected costs if you are still in the process of planning your venture.

By not setting prices that are too high or too low for the products you are selling, you will be assured of a favorable position in the market and a healthy share of the wealth.

And remember, the most important variables to take into account when developing a price structure are

1. Your costs

2. Your expected profit

Success Tips for Your Coffee and Tea Store

1. Take the time to develop new and unique products to offer your customers.

2. Maintain the highest standards in everything you do; it will ensure a solid reputation.

3. Never sacrifice quality as a means of saving money or time.

4. Plan ahead. Allow yourself time to work efficiently. The bonus is that you will enjoy it a lot more.

5. Be creative.

6. Make customer service the byword.

Prices for Your Coffees & Teas

Knowing that the industry is growing, you will probably want to base your prices on the following:

1. Purchase price

2. Gourmet value of particular blend

3. Weight

4. Cost of associated labor (for example, time spent brewing house blend or creating a gift basket)

5. Competitors' prices

6. "Extras" provided (a special item you procured for a customer)

7. Quantity discount

8. Club member or frequent purchaser discount

The idea is to stay financially alive in your business. In order to keep it going, you may want to figure out how much coffee and/or tea you need to sell in a day, in a week, and on a monthly basis in order to make the amount of money you feel is necessary to survive as an entrepreneur.

Pricing According to Customer Value

According to a specialty food consumer report, gourmet coffee consumers are affluent, earning over $35,000 per year. College graduates consume 49 percent more coffee than average; those with some college education consume 11 percent more than average. The majority of gourmet coffee consumers are concentrated in large urban communities.

Your customers will tell you what they want to see in your store. They are the best source of information as to what you should carry. Ask your customers if they are happy with your products, prices, and services.

Coffee and Tea Gift Baskets

Creating coffee and tea gift baskets could be a big boon to your business, offering an added specialty that other gourmet coffee and tea stores do not have. Average prices for baskets run between $35 and $60, with variations for location and special items. The prices you set for each of your baskets is primarily up to you; however, they must be in line with what the market will bear. The price you decide upon will probably be based on the following:

1. What you spend for the basket and its contents
2. Number of items included
3. Type of items included
4. Size of the basket
5. Time spent creating basket
6. What other stores are charging
7. Amount of expertise you have and your reputation in the community
8. "Extras"—if any—you are providing (perhaps some special item you went out of your way to get for a customer)
9. Quantity of baskets a customer is purchasing at one time—you may want to offer a quantity discount
10. Number of handmade items included
11. Time and type of delivery

Your customers will tell you what they want to see in the store and, as a result, they are the best source of information you have as to what you should carry. You can cater to the price range of your customers by asking them how much money they want to spend on a particular basket, and then, based on your knowledge of income needs and expenses, you can tell them what, for example, a $25.00 gift basket will include.

They Want to See What They Pay For

If your basket has a price tag of $50.00, it better look like it's worth $50.00. And if you do special order an item, you want to be certain

that your customer is going to purchase that item—especially if it's something you might have trouble selling to someone else.

It is a good idea, therefore, to ask for at least partial payment on a basket at the time it is ordered so that if a customer has a change of heart and fails to pick up the basket, you are at least reimbursed for some of your expenses.

When people pay a down payment on an intended purchase, they are more likely to follow through. Establish a standard down payment according to the effort and time you will be putting into creating baskets. Although the possibility is good that you will be able to sell a completed basket "off the shelf" at a reduced price if the original customer doesn't pick it up, a 25 percent surcharge will help cover expenses.

The Mail-Order Coffee/Tea Sale

Another way to increase sales is to offer direct-mail coffees and teas to those people who live in areas where they can't buy them directly. This possibility was mentioned previously, but it is included again to remind you of its merits for extra-income potential. You want to offer direct-mail customers good prices, but since they are a captive audience, you can price for good profits. A markup of from 30 to 40 percent is considered normal.

It might seem odd that people would be interested in purchasing a staple product through the mail; however, mail order coffee sales are a fast-rising category within the industry. Some specialty coffee vendors, such as Gevalia and Green Mountain Coffee, sell their products exclusively through mail order. Many others start with a single store location and expand into the mail order business as they add locations privately or through franchising. You may start by limiting your initial offerings to top-quality coffees and teas and including other merchandise, such as t-shirts emblazoned with your coffee house's logo, as your mail order business grows.

A word of caution: There can be major expenses associated with establishing a mail order business. Significant expenses include name acquisition through purchases of mailing lists, catalog production, and mailing fees. As discussed in Chapter 6, if you invest in a computer and laser printer, you can save some costs by preparing your own advertising and marketing materials.

If you can develop a strong customer base, profit margins in mail order are high.

Franchising

Several coffee houses have expanded through franchising and/or licensing. Franchising offers an entrepreneur new capital and people to expand an operation. In the specialty coffee retail industry, the major national franchises include Barnies Coffee & Tea Company, The Coffee Beanery, Coffee Connection, Gloria Jeans's Coffee Bean, and Starbucks, Inc. In most regions of the country, franchising conventions are held several times a year. You may attend a convention and learn about franchise opportunities in this fashion, or you may contact the specific franchise that you are interested in directly.

Product Branching

In addition to specialty coffees, teas, and associated products, most coffee houses carry a variety of desserts and pastries. To even out sales throughout the day and to account for seasonal variations, many stores have branched out into several other areas. Some typical sidelines include offering a light food menu, including cold coffee products and frozen confections during warm months, offering music and entertainment in the evenings, and merchandising t-shirts, mugs, and other items with the coffee house's logo. Many coffee houses are linked to bookstores, since a comfortable place to drink coffee and read is a natural combination. The associated products and services you offer depend upon your finances.

Initially, there will be a limit as to how much inventory you will be able to finance. As your business grows, however, you will be able to add items that you know will answer customer demand and acceptance. If you shop carefully, you will be able to find items that will enhance what you are already selling in your shop as well as selling well on their own.

Notes

Key Points

Personal Thoughts

Additional Research

17

INVENTORY AS INVESTMENT

\mathbf{A}sk 100 small business owners what inventory means to them and more than 90 percent will tell you it's the merchandise they keep on hand to sell to their customers, or the materials and supplies stocked to produce goods or perform a service.

This is somewhat accurate, for inventory can and should be viewed as any supplies, raw materials, or finished goods used to generate a profit in your business. But it isn't the response that a savvy business owner would give.

Surprisingly, according to a recent study conducted by a leading consulting firm, less than 10 percent of a group of 500 entrepreneurs interviewed spoke of their inventory in terms of the *investment it represents;* an investment that can range from 15 to 25 percent of total operating capital.

It is because of this misunderstanding that many small business owners often fail to incorporate good inventory control practices into their regular management routine. Although they keep an eagle eye on every penny going through the books, they may totally overlook the cash tied up in their inventory.

Controlling Inventory

Inventory control can be a very simple, straightforward task if you implement a workable system from the beginning—preferably even *before* you start ordering and receiving goods. You will find that time really flies when you are self-employed, and it's very easy to postpone such tasks as inventory control until the inevitable day you find yourself facing this overwhelming job. Inventory control will give you valuable information about: (a) whether or not you are carrying too much or too little inventory based on, for example, items and prices preferred by your customers, or seasonal aspects, and (b) if you are realizing optimum economy determined by the costs of storage, taxes, handling, and the investment per unit.

The ideal situation is to maintain an inventory that is profitable because it turns over (comes in and goes out of the business) regularly, thereby lowering the cost of storing, displaying, and insuring it.

Although there are countless alumni of the school of hard knocks, there has not yet been a move to accredit that institution.

Sonya Rudikoff

There are several methods of inventory control that you can adopt, depending on your business. The main goal with each method, however, is to tell you how many items you have on hand and how many you need to meet your customer or production demands. It will also work toward lessening inventory shrinkage, which is generally the result of employee pilferage; customer theft; incorrect inventory storage; or sloppy record maintenance of items ordered, received, and used.

You can tell how many items you currently have by making an educated guess, which generally only works for businesses having a small, visible inventory that is relatively predictable. An example of the kind of business that could probably operate efficiently with this relaxed form of inventory control would be a

one- or two-person enterprise that monthly goes through a box of invoices and similar supplies available for a minimum amount at the discount office-supply store.

Other methods of inventory control are *the physical count,* which should be done at least once a year for tax purposes, or—the easiest and most efficient of them all—*an ongoing* record. Ideally, you should incorporate both systems by backing up periodic physical counts with a perpetual record.

To set up your perpetual system, simply create a file card or inventory sheet in a three-ring binder for each item in your inventory. Across the top of the card or sheet, list the following:

1. Item name and code number, if applicable
2. A description of the item
3. Where it is stored
4. The supplier's name, address, and phone number
5. Unit price (i.e., $12.95/dozen)
6. Your selling price if a retail item, or percentage of gross price of completed product, if used for manufacturing or service
7. The date you place an order
8. The number of items and the date they are received

Then, every time you sell or use an item, write it down and subtract it from the last balance. You should also indicate the reorder number, based on when and by how much you must replenish your stock.

The reorder number will be determined by such factors as the following:

1. The minimum cost per unit available from your supplier, including quantity discounts, preseason specials, and discounts for cash or quick payment
2. The delivery schedule—the delay between placing the order and receiving it
3. Economic and social trends that can affect the way an item is perceived by the public

For example, during a depressed or inflated economy, sales for leisure items typically drop. By keeping an eye on these factors, you can adjust your inventory needs accordingly and not get stuck with great quantities of items that you can't move.

After a while, you will be able to recognize at a glance which items are movers and which are simply taking up shelf space. When you reach this point, your ordering skills will become much more efficient and your investment in inventory will become a profitable proposition.

Coffee and Tea Store Inventory

Since you will want to carry a good selection of items, you will need to buy from several different sources and wholesalers. As a result, it is helpful to find one company that carries many of the items that you will use regularly. The fewer companies you deal with, the less work for you.

Almost half of the specialty coffee stores surveyed indicated that they purchase coffee from suppliers once a week and most do not do their own roasting. Nine out of ten owners sell coffee in bulk form, which represents 70 percent of store sales, and one out of two stores is also likely to carry its own private label blend. More than half sell brewed coffee by the cup.

According to studies conducted by The Specialty Coffee Association of America, specialty coffee stores designate 58 percent of their space to coffee-related items, 35 percent of which is devoted to beans.

Promoting your business as a gift store as well as a coffee and tea store gives your customers more choices. Also, by keeping a selection of cards and small items on display at all times, you increase the chances of attracting browsers who may turn into coffee and tea store clients.

The intellect has little to do on the road to discovery.
There comes a leap in consciousness, call it intuition or
what you will, and the solution comes to you and you
don't know how or why.

<div align="right">Albert Einstein</div>

All of the coffee and tea store stores we visited carry several lines of greeting cards, a natural addition to inventory. The majority also branch out into the gift area by incorporating displays of small knickknacks and home accessories into the decorating scheme. Part of the fun of setting up your store is in locating unusual items and making a personal statement based on your sense of style and taste.

You must also decide how much display inventory you are willing and financially able to carry. In the beginning, your choices may be limited by overhead demands, but as the business grows, you can add gift items and products according to customer demand and acceptance.

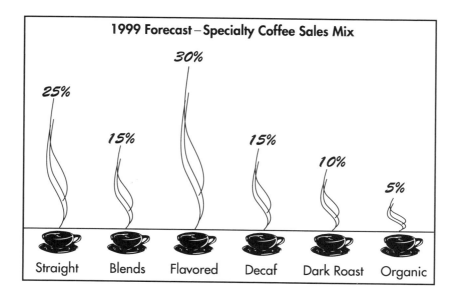

A regular inventory check of the products you have available is an absolute must. If there's one thing that's bad for business, it's running out of something that a customer wants. You will need to know how much of an item to order and how long it will take for delivery to ensure receipt before you need that item again. Saturdays are by far the busiest day of the week for stores, followed by Fridays, then Thursdays; December, followed by November, comprise the busiest months. July and August, somewhat logically, are the slowest months.

Know your customers and your store's patterns. Develop a relationship with suppliers so that you are assured of getting deliveries on time. After doing this over and over again, you will develop your own rhythm in performing all of these tasks, and they will become second nature to you.

Notes

Key Points

Personal Thoughts

Additional Research

18

MAKING THE MOST OF YOUR TIME

A recent survey of small business owners indicated that one of the qualities they felt contributed the most to their success was organization. In conjunction with this is the fact that time management and basic organizational seminars continue to be the most popular offerings in adult education catalogs and business workshops around the country.

Time is money! It is vital to your success that you learn to manage your time and organize paperwork. This might sound rather simplistic, but you would be amazed at the number of small business owners who operate in a constant state of chaos.

Although we have seen a number of "A messy office is the sign of a creative mind" posters on entrepreneurs' office walls, it is a good bet that the holders of these signs can recount story after story of missing checks, lost orders, or misplaced files, which disrupted or halted the flow of business until they were located.

The survey respondents also stated that once they had learned to manage their time in both their personal and business lives and had set up guidelines for handling routine tasks, they felt more confident about accepting new challenges and making decisions.

The simple truth about getting organized is that it clears your mind for taking care of the nitty-gritty, profit-making aspects of

being in business—production and promotion. For example, just by setting aside a certain place in your work area to hold customer files and billing information, you have made one major step toward maximizing production. Knowing that all the supplies and materials needed to conduct your business are located in one place saves you both time and energy.

Making Time Work for You

Time management is the ability to take the hours we have available and use them to our advantage. Making lists of tasks to be done and giving them a priority rating is one of the best ways to avoid losing precious moments.

Keep a monthly calendar handy to help you keep track of major commitments, important dates, and appointments. Try to avoid using it for notations of daily work in progress, carryover tasks, or other things that are best suited for inclusion on your daily and weekly lists. There are several other ideas you can easily incorporate into your working lifestyle that will maximize your productivity.

Work Smart

Handle the jobs you find most difficult or cumbersome during peak performance time. If you are a morning person and find that making telephone calls is one of your least favorite tasks, make those calls first thing in the morning when you are feeling fresh and energetic. Then organize the rest of the day's tasks accordingly.

Set Realistic Daily Goals for Yourself

Just because you are the key employee, don't try to do everything at once. Learn just how much work you can accept and expect to accomplish in a given period of time and allow yourself to turn down work if it seems like it will be too much for you to handle.

Positive Reinforcement

Many small business owners work alone or with a limited staff when getting their businesses off the ground, so there generally

aren't many people around to support or praise your work. Everyone needs strokes! Remember that a customer's praises are proof that you are doing a good job.

Reward Yourself

Sometimes it is easy to forget that you are important to the business—you *are* the business. So treat yourself to a special dinner now and then or plan an evening where you go out to a movie or do something else that has absolutely nothing to do with business. Remind yourself that this is your reward for accomplishing certain goals for your business. It will help to keep your spirits high.

Don't Procrastinate

Don't put off doing tasks that must be done. If you constantly let some tasks slide because you don't enjoy doing them or they are boring and time-consuming, you will soon find yourself backlogged and unable to catch up. The effects of procrastination may not be obvious until you are faced with a rush in business and, at that point, you will realize that you are working at less than maximum efficiency using the wrong supplies, feeling tense, and being hassled by minor problems that could have been avoided.

Limit Personal Phone Calls

Ask friends and relatives not to call during established business hours. These friendly calls not only eat into productive time, they tie up the line when a cusstomer may be trying to get in touch with you. The same holds true with visitors. If you are self-employed, friends and family often feel that they can stop by anytime to chat. Explain that you will be happy to visit with them at some other time and be sure to tell them why so there aren't any ill feelings.

Delegate

If you find that you absolutely cannot handle a certain aspect of the business, such as your own bookkeeping, for example, don't labor over the task—you will end up wasting a great deal of time and could make some serious mistakes. Admit to yourself that the

task is just not your strong point and make arrangements to have someone else do it.

Learn to Say "No"

One of the hardest things for most people to do is to say "no." Even when we realize that helping a friend out on a special project will cost valuable time, we often agree to do it because we hate to say no. What we do is justify our acceptance by assuring ourselves that saying yes will put us in the position of meeting a lot of potential customers. The reality is that using that time to perform some task related to your own business will probably result in a paying customer, not just a potential contact.

Plan Your Time

When you have to run errands, plan them for a sensible time, such as in the morning when traffic is light. Plan to do as many things as possible in one trip, outlining the stops you must make so they follow a sequential order.

Notes

Key Points

Personal Thoughts

Additional Research

19

COFFEE AND TEA PRIMER

Knowing Your Coffee Products

Why should your customers pay more to buy coffee from your store when they can buy ready-made ground coffee from the supermarket for much less? You must be able to answer that question, whether you are convincing a customer or a bank loan officer.

For starters, jars of instant and cans of ground commercial coffees are different from specialty coffees that come as whole beans and need to be ground before brewing. Commercial coffee is usually roasted and packed in large plants, under nationally advertised brand names. "Gourmet" coffee is usually roasted in small stores or factories using traditional methods and technologies and sometimes is sold right where it's roasted.

Specialty coffees are typically sold in one of three ways: (1) by the place where the bean originated, such as Colombia or Kenya, (2) as a particular blend, such as "Mocha Java," or (3) as a special kind of roast, such as "French roast" or "Italian roast." Commercial coffee, although sometimes labeled "100 percent Colombian," is more often sold simply as "coffee." In addition, many of the grocery store's instant coffee mixes with exotic-sounding names are most definitely not specialty coffees.

Flavor and aroma are the most important differences between commercial and specialty coffees. Specialty coffees are fragrant and come in a wide variety of flavors. Coffee beans are divided into two broad categories: robusta (ro-boo-sta) and arabica (a-ra-bee-ka). Arabica, the more delicate and flavorful of the two types, offers a balanced aroma and a sweet, acidic taste. Arabica coffee represents 75 percent of the world's coffee production. Robusta coffee has a stronger flavor and heavier body and twice the caffeine content of arabica. Robusta coffee represents about 25 percent of the world's production. Many supermarkets carry blends of robusta and arabica beans, but gourmet and specialty coffee stores carry only arabica beans.

Two processing methods exist: the wet (washed) method and the dry (unwashed) method. Wet-processed coffees command higher prices because of the additional labor and equipment required for processing; color ranges from blue/green to gray/green. Dry-processed coffees are generally green to brown in color.

Coffee is graded according to a number of factors: size, shape, density (hardness or softness), color (from blue to brown), and physical defects (number of broken beans or detritus found in a given sample). Each country sets its own standards for sorting, grading, and labeling coffee.

Robusta coffees are grown between sea level and 2,000 feet and are processed using dry methods. Arabica coffees are grown in altitudes nearly 6,000 feet above sea level and are processed using wet methods. Arabicas are the finest coffees grown in the world. They are known for their unique full-bodied flavor, rich aroma, density, even roast, and balanced flavor. They are special because of how and where they are grown and the labor-intensive handling the coffee bean receives. It takes a whole year for a coffee tree that produces arabica beans to produce one to one and a half pounds of coffee.

Purchased fresh and brewed correctly, gourmet coffees are superb. It's good to remind customers who pay about $7.50 per pound for a specialty coffee that each cup costs about 10 cents—far less than they pay for a cup of very standard coffee in a restaurant.

Coffee should be black as hell, strong as death, and sweet as love.

Turkish proverb

Coffee, with a value of approximately $10 billion annually, is the largest commodity, next to oil, in international trade. More than ten billion pounds are grown in fifty countries providing more than twenty million jobs.

Specialty coffee stores generally fall into one of the following categories:

1. Roasts their own coffee on the premises or nearby

2. Part of a chain, roasts their own coffee at a central location

3. Buys their coffee from another local specialty roaster

4. Buys their coffee from several roasters

5. Buys their coffee green and has it roasted for them

You should make every effort to buy your coffee as close to the person who does the roasting as possible, since the coffee will be fresher and the salesperson probably more informed.

It is possible to brew coffee badly, so learn how to do it well. Brew at least two different kinds of fresh coffee for your customers to sample and rotate choices to keep the customers interested. If you brew your coffee in the morning and keep it in a thermos all day, it will gradually lose whatever flavor it had to start.

Coffee Names

All coffee names refer to the degree to which the bean is roasted, the place the bean came from before it was roasted, or the dealer's name for a particular blend of beans. Darker coffees are given European names and are distinguished by the length of time the bean is roasted.

Italian roast, for example, is darker in color and has been roasted longer than Colombian. These names do not refer to the

origin of the coffee beans. Sometimes, however, coffee sellers specify the coffee that they have dark-roasted, thus the name "Dark Roast Colombian." In most cases, a darker-roasted coffee will be an inexpensive blend from a variety of countries, and the name will refer only to the roast.

Lighter coffees are roasted to what are commonly known as typical "American" tastes. The difference is not the roast, but the origin of the bean itself. These coffees carry exotic names, like "Sumatran," probably because most can only be grown success-fully in or very near the tropics.

Sometimes coffees come as blends. There are two reasons to blend beans. One is to make a coffee with a flavor that is better, or different, from the flavor produced by "straight" coffee. The other is to cut costs. Commercial coffee blenders know exactly how much cheap robusta they can get away with without sacrificing a lot of flavor.

Mocha Java is a blend of mocha and java beans. There are high-priced blends and low-priced blends, morning blends and after-dinner blends, dark-roasted and light-roasted blends. We suggest that you get to know your blends and straights very well by doing some of your own research, reading and tasting as much as you can, and talking to specialty coffee wholesalers. The more you know about your product, the more competent you will be to sell it.

Tasting Coffee

Experienced coffee tasters can distinguish the origin of most coffees in a blend simply by smelling and tasting. Since descrip-tions of the differences in coffee tastes are difficult to put into words, you will need to savor the various choices and learn to be able to offer a customer an idea of how each coffee tastes.

The little brown bag you put the coffee beans into says nothing about the beans the way a wine bottle discusses the body of the wine it contains, so any explanation you can give your customers will help. Many people do not understand the language of coffee and will rely upon your description and judgment.

The following is a list of the main families of coffee.

Arabian and African Family

Ethiopian Harrar
Kenya
Tanzania
Yemen Mocha

Known for their distinctive winy acidity.

Individualists

Hawaiian Kona
Indian Mysore

Indonesian & New Guinea Family

Celebes or Sulawesi Kalossi
China Yunnan
Java Arabica
New Guinea
Sumatran Mandheling or Ankola

Heavy bodied with a lower-toned, richer acidity.

Latin American Family

Brazilian Bourbon Santos
Colombian
Costa Rican
Dominican Republic or Santo Domingo
Ecuador
El Salvador
Guatemalan Antigua
Haitian
Jamaican Blue Mountain
Mexican Coatepec
Nicaraguan
Oaxaca Pluma
Peruvian Chanchamayo
Venezuelan Maracaibo

Characterized by a bright brisk acidity and clean straightforward taste.

The following qualities are important for you to be able to distinguish in your coffee tasting.

1. *Acidity.* This is a pleasant tartness, a dryness the coffee leaves at the back of your palate and under your tongue.

2. *Body.* This is the sense of heaviness, richness, thickness at the back of the tongue when you swish the coffee around in your mouth. The coffee is not really heavy but simply tastes that way.

3. *Aroma.* Acidic coffees smell acidic and richly flavored coffees smell richly flavored. It's that simple.

4. *Flavor.* Acidity, body, and aroma all have to do with a coffee's flavor. Some coffees have a fuller, richer flavor than others, whereas some have an acidic tang that dominates. Some coffees have a distinctive flavor whose characteristics just plain stand out. A flavor can be mild, mellow, or harsh.

Blending Coffee

Roasters blend coffees to

1. Create a new flavor

2. Cut caffeine content

3. Cut costs but maintain quality

4. Achieve a better taste (the best reason to blend)

You can learn to blend and roast green coffee and make your own special blends. This will save you money and offer you the opportunity to come up with a unique blend that your customers will come only to you for.

The most famous blend, which combines Arabian Mocha with Indonesian Java began when Mocha and Java were the only two coffees available. Mocha is a mildly acid, winy coffee with a fairly light body. Java is a heavier-bodied coffee that is sweeter and deeper toned. Together they make a brew that is less distinctive than either one but more balanced.

To create your own personal blend, you will have to learn to distinguish among acidity, body, and flavor. If you have a favorite coffee that seems slightly less than perfect to you, combine it with the opposite extreme of what your favorite lacks and see what you come up with.

Roasting Coffee

The coffee bean loses a lot of moisture in the roasting process, so it weighs less after roasting. It loses some protein, 10 to 15 percent of its caffeine, and traces of other chemicals. Sugars are burned or caramelized, which contributes color and body to the cup.

There is such a thing as not roasting coffee long enough or hot enough, or roasting too long or at too high a temperature, so you have to know what you're doing. Check the "Resource" section of this manual for books, magazines, and other sources that will help you learn how to roast coffee beans on your own.

Advantages of roasting your own coffee include (1) satisfaction, (2) fresher coffee, and (3) significant cash savings.

You should be able to get green beans for about two-thirds of what you would pay for roasted beans. You might be able to get a further break in price if you buy in quantity. Before you go for a bulk purchase, however, try roasting your own to determine whether or not it's something you want to do.

If you live in a town or area full of coffee lovers who can't obtain good gourmet coffee, your roast could be a big seller. You can get a good price on a big sack of green coffee beans, drink much better coffee than people who frequent city coffee shops, and cut down on expenses.

Grinding Coffee

Whether or not you decide to make your own private roast, you will need to know about grinding coffee because this is something you will definitely be doing in your store. Ground coffee begins to lose its flavor an hour after grinding, so you will always want to grind your coffee beans right before selling or brewing.

Advise your customers that their coffee will stay fresher if they buy whole beans. Grinding one's own coffee takes little time,

and grinders are inexpensive. These days there are a lot of quick electric varieties that make it a snap.

Grinding fresh coffee beans is the best practice for increasing the quality of coffee. Other tips for maximizing the flavor of coffee include using an airtight storage container and freezing coffee beans for long-term storage. For short-term storage, keep coffee in a cool, dark place.

There are four ways to grind coffee.

1. *Mortar and Pestle.* This grinder is cheap and keeps your wrist muscles in shape. If you have a lot of time and don't mind that you can't grind much at a time, it's the perfect grinder.

2. *Millstone.* Used to reduce grains to flour, the millstone creates a uniform grind, although it, too, is slow. It's a hand grinder, like the wooden hand-held models through which beans are fed and then ground into coffee that falls into a drawer at the bottom.

3. *The Roller Mill.* This is used only in huge commercial grinders.

4. *Electric Blade Grinder.* This is the grinder of today. It does not make a uniform grind, and some coffee turns to powder, but it's quick, cheap, and doesn't take up much space. If you have room in your store, you should carry these for sale; customers who are fanatic coffee drinkers will want to own one if they don't already. They will be good sellers for incurable coffee lovers and are also a good additional gift accompaniment to a pound or two of coffee.

Brewing Your Coffee

The only equipment you need to make great coffee is an open pot, a flame, and perhaps a strainer. The best pot, however, is made of glass or porcelain. Stainless steel is okay, but aluminum is not. You want a pot that comes apart easily for cleaning and one that has no seams or cracks. On occasion, you should run a strong

solution of vinegar through your brewer in order to clean parts that are prone to deposits and build-up.

The best brewing rules are listed below.

1. Use fresh coffee, ground as close to the moment of brew as possible.

2. Grind coffee as fine as you can make it but never to a powder.

3. Use plenty of coffee—at least two level tablespoons or one standard "coffee scoop" per five- to six-ounce cup.

4. Keep your coffee maker clean and rinse it with hot water before you brew.

5. Use fresh water, as free of impurities and alkalines as possible.

6. Use hot water, as opposed to boiling or lukewarm water. The ideal water temperature is 200 degrees. A good way to go is to bring the water to a boil and wait two minutes before pouring over the coffee. You don't ever want to boil coffee, since boiling removes all of the delicate flavoring. Perking and reheating coffee are definitely not advised, since they will also destroy the flavor.

7. Ninety-nine percent of a cup of coffee is water. Use bottled or purified water.

If you live in a "hard" water area, you can compensate by buying more acidic coffees (African, Arabian, and the best Indonesian and Central American beans) or brew with bottled water.

Methods of Brewing

There are a variety of brewing methods. The main methods are listed below.

1. *Open Pot.* This is the simplest method and as good as any of the other options. Steep the ground coffee in a pot of hot water, strain the grounds from the brewed coffee, and serve.

2. **Plunger or French Press.** The pot is a tall cylinder. A fine-meshed screen plunger fits tightly inside the cylinder. You put the ground coffee in the cylinder, pour boiling water over it, and insert the plunger in the top of the cylinder without pushing it down. After four minutes, the coffee will be steeped. You push the plunger down through the coffee, forcing the grounds to the bottom of the pot and serve the coffee directly from the cylinder. This plunger is also used for steeping teas.

3. **Drip.** The drip maker has two compartments, divided by a metal or ceramic "filter" or strainer. The ground coffee is placed in an upper compartment, and hot water is poured over it. The brewed coffee trickles through the strainer into the lower compartment.

4. **Percolator.** You won't find these in a coffee specialty store because they boil coffee and remove aroma and flavor from coffee.

5. **Vacuum.** This method does not differ greatly from coffee made with a filter cone. It was popular in the 1920s and 1930s and is now limited in production. It's a very pretty but not a quick or efficient way to brew coffee.

6. **Filter.** A paper filter, usually cone-shaped, is placed in a plastic, glass, or ceramic holder. Finely ground coffee goes into the filter, which goes on top of a glass or ceramic flask. Hot water is poured over the coffee into the container. The filter's advantages are a clear brew with a minimum of oils and sediment. Paper filters make grounds easy to dispose of and the coffee maker easier to keep clean.

7. **Automatic Filter Drip.** This coffee brewer heats water to the optimum heat for coffee brewing and automatically measures it into the filter. The coffee drips into the decanter and an element under the decanter keeps the coffee hot, once brewed. The coffee drinker measures cold water into the top of the maker, measures coffee into the filter, presses a switch, and in four to eight minutes has four to twelve cups of coffee. Although this system does

not make for the best coffee, it produces a coffee superior in flavor to percolated coffee.

8. ***Concentrate Methods.*** To make a hot water concentrate, use eight cups of water to a pound of finely ground coffee and brew in whatever way you like. You can halve the recipe or brew twice if your coffee maker won't handle a pound at a time. Store the resulting concentrate in a stoppered bottle in the refrigerator. To use: In a preheated cup, add about one ounce to every five ounces of hot water.

 The cold water method is to steep a pound of regular grind coffee in eight cups of cold water from ten to twenty hours, filtering the resulting concentrate into a separate container. Store it in the refrigerator. When ready to use, add an ounce of concentrate to a cup of hot water.

 The result of the concentrate method is a very mild, delicate brew with little acidity, light body, and a natural sweetness.

Experimenting with various brewing methods will be essential for you, especially if you decide to sell different brewers in your store. Your customers will want to know the advantages and disadvantages of each, and a good coffee specialty store owner should have the answers.

Espresso

If you have to choose only one kind of coffee to sell, this is it. Most specialty coffee stores sell lots more than one kind, but if you are vending coffee from a pushcart, and your selection is limited, espresso is the winner by far. It is a unique method of coffee brewing in which hot water is forced under pressure through coffee, a cup at a time.

Espresso is a dark brown roast, not quite black. It is a unique approach to coffee cuisine, involving not only roasting and brewing, but grind and grinder, a technique of heating milk, and an entire menu of drinks. It carries with it a sense of mystique, which explains why it is the centerpiece of the European style of cafe.

There are home espresso machines that you might want to stock in your store if you decide to sell coffee makers. The best espresso and cappuccino come from the big bulky machines used in the finest specialty coffee locations.

Espresso machines are like any brewer in that they steep ground coffee in hot water, but in normal brewing, the water seeps by gravity down through ground coffee resting in a filter. In the espresso process, the water is forced under pressure through finely ground coffee packed tightly over a filter. It's the pressurized water that makes for the great taste. The water moves so quickly through the coffee that there is no time for unpleasant chemicals to be extracted.

An added advantage is that every cup of espresso is brewed instantly right before your eyes, seconds before you drink it. Early in the history of the espresso machine, someone discovered that the steam collected in the top of the machine tank could be used to heat milk as well. Thus, a new aspect of espresso making was born. Add a dash of cocoa or cinnamon to the steamed milk in the espresso, and you have a tasty delight that can be served up to look either funky or classy.

Espresso Drinks

1. ***Espresso.*** A single demitasse of black espresso coffee, sometimes served with sugar or a drop of a flavored extract such as vanilla or almond

2. ***Espresso Romano.*** A demitasse of espresso served with a twist of lemon on the side

3. ***Double or Doppio.*** A six-ounce wallop of a cup of espresso

4. ***Cappuccino.*** One-third espresso, one-third hot milk, topped with one-third foam, in a heavy cup, usually with added sugar

5. ***Cafe Latte.*** One or two shots of espresso and three times as much foamed milk, served in a large cup or glass; considered a breakfast drink in southern Europe and Latin America

6. ***Espresso Macchiato.*** A demitasse of espresso "marked" with a small quantity of hot, foamed milk

7. ***Latte Macchiato.*** An eight- or ten-ounce glass of half hot, foamed milk, into which a demitasse of espresso is slowly dribbled creating layers of color shading, darker toward the top, with white foam at the top

8. ***Mocha.*** One-third espresso, one-third strong, un-sweetened hot chocolate, and one-third steamed milk

Tea

While there are specialty stores carrying only coffee, we don't know of any that carry only teas. Usually gourmet coffee stores will carry special teas as well. When tea was first sold, before the invention of the tea bag, a blend was made up by a grocer and sold over the counter. This form of custom buying still exists in specialty stores, where loose teas are showcased in tightly covered jars on a counter or store in labeled tins.

Most teas sold are not originals but blends because it is simpler to maintain a consistency in blended teas. The blends can be mixtures from several gardens in the same district, such as Darjeeling, Assam, Ceylon, or China. Or they can be mixtures from different districts (or different countries), such as Earl Grey, English Breakfast, or Irish Breakfast.

The evergreen tropical plant from which the tea leaf comes, which can grow up into a 30- to 50-foot-high tree, is cut back to no higher than four feet to encourage the continual growth of young shoots. These shoots ("flushes") produce two tender young leaves and a small bud used in tea manufacturing. The picking of shoots is done by hand. An accomplished plucker can pick almost forty-five pounds of shoots per day, enough to become ten pounds of salable tea. The baskets of shoots are taken to the factory to be processed. The green leaves then go through the following stages:

1. ***Withering.*** Leaves are spread on shelves and wither naturally or are heated to remove as much water content as possible.

2. **Rolling.** The leaves go through rolling machinery that breaks up leaf cells and frees the juices essential to flavor. Fine leaves dropped through mesh sieves are gathered up and taken to a fermenting room.

3. **Fermentation.** The leaves continue to absorb oxygen, as they turn the color of new copper. This process distinguishes the black teas most Americans drink from green teas bought by Moslems, whose religion forbids the use of any fermented product.

4. **Firing.** This drying process halts further oxidation of the leaves by sending continuous blasts of hot air onto them.

Once the leaves have gone through the drier, they are ready to be graded. They take on one of several names that determine the size and appearance of the leaf but have nothing to do with the tea's quality.

Leaf Grades

There are two main grades: leaf and broken leaf. Leaf grades are large and include orange pekoe, pekoe, and pekoe souchong. The broken leaf grade includes broken pekoe, broken orange pekoe, broken pekoe souchong, fannings, and dusts. These last two are the smallest-sized leaves and are valued for their strength and quick infusion. Broken leaves give a darker liquor with a stronger character than the leaf. This generally yields more flavor and fragrance.

Green tea is a bit different. It is usually plucked without stalk and never withered or fermented, the leaf remaining green. Oolong, the best quality grown in Taiwan, is half black and half green. It is slightly withered and then steamed to make the leaf pliable. The leaves are rolled and fired, coming out greenish-brown with a peachlike flavor.

Pouchong, little known to the general public but preferred by the Chinese, is neither green nor an oolong, and is slightly greener. Many pouchongs are scented with flowers.

Tea Auctions

Of the six tea auction centers around the world, five are called primary auctions, and one is called in-transit. The primary centers are Calcutta, for northern India; Cochin for southern India; Colombo, Sri Lanka; Jakarta, Indonesia; and Mombassa, Kenya, for east Africa. London, the in-transit market center, receives teas on consignment that are sent from all over and cataloged for sale.

Once tea chests are stored in special warehouses, weighed, and marked, a small hole is bored in each and samples of the teas are made available to tea brokers, who send smaller amounts to clients all over the world. The brokers have their own tea tasters who evaluate samples before placing their orders. The amount of tea that can be sold in one day is staggering—five million pounds is not uncommon.

A broker will sometimes put a value on a certain chest, taking into account (1) its weight, (2) the time of year, and (3) how the tea was packed. Only specific operators can participate in the auctions, but buyers' representatives are nearly always present just to watch or communicate with their brokers.

These days, bidding on a lot of tea can take place in a matter of seconds with the lift of a finger. Once the sale of tea is completed, the auctioneer announces the name of the buyer, how much tea was bought, and how much it cost. He also provides everyone with a catalog noting names and figures.

The Blending of Tea

Tea blenders are responsible for producing a uniform taste for each blend sold, a flavor that will not change from year to year regardless of whatever weather has befallen the tea plants. The highly professional blender or taster can tell where leaves were grown simply by handling them. Sometimes blends consist of two or three teas. Other blends will be a mixture of up to fifteen teas. It is easy to understand how, in a less expensive blend, one or two teas could be removed from a fifteen-blend tea without a noticeable change in flavor.

Packaging Tea

1. ***Tea Bags.*** Tea bags developed by accident in 1904 when a tea merchant, anxious to acquire new customers, sent out samples of his teas wrapped in small silk bags. A few of his customers put the bags in cups or pots, pouring hot water over them. Today's bags are not made of silk but of a specially treated filter paper. A single machine assembles the tea bag, measures the prescribed amount, fills it, seals it, and attaches a string and tag.

2. ***Loose Tea.*** Loose tea is handled by a bank of machines that take a piece of cardboard, shape it into a package, line it with paper, fill it with measured loose tea, seal and label it, and place it on the conveyor belt.

More expensive lines of tea can be bought in decorative canisters or custom-made boxes. Gourmet tea lovers, however, will go for pure or blended teas in a simple plastic bag.

Once you have loose tea and some hot water, the process of tea making is simpler than that of coffee brewing. It's simply a matter of how strong you like your tea. If you like it strong and flavorful, you leave the tea leaves in the water longer, with a tea cozy on the pot.

Herbal Teas

Early coffee brewers believed that coffee had medicinal qualities. It was sometimes prescribed for intestinal disorders and hysteria. Now, because of its caffeine content, the medical community is suggesting limited drinking of coffee.

Teas contain almost as much caffeine as coffees, but herbal teas have no caffeine. These are teas that are believed to have medicinal properties. To this day they are sold in health food stores. Books have been written on herbs and which teas to drink for certain maladies.

There is a strong distinction made between customary teas of the world and herbal teas. While teas like Earl Grey are sold in specialty food and gourmet coffee stores, herbal teas are generally found only in health food stores and supermarkets.

Notes

Key Points

Personal Thoughts

Additional Research

20

HIRING EMPLOYEES

There will come a point when you find you simply cannot do everything by yourself. You will need to consider hiring an employee to help you. Although your profit margin will be higher if you do it all alone, when business reaches the point where you are not able to handle all the customers in addition to maintaining the store and taking care of inventory control, it's time to get help.

You can begin by getting someone on a part-time basis. That person may be interested in working full-time when the need arises, but you can play that by ear. In the meantime, you know that for a certain number of hours per day or per week, someone else will be there to do those things you are simply too busy to get to. The idea is to keep the customer happy, and if that means hiring another person, that's what you do.

Knowledge is of two kinds. We know a subject ourselves or we know where we can find information on it.

Samuel Johnson

Good Pay for the Right Person

For the absentee owner, the cost of a good manager must be figured as an additional, but vital operating expense. Experience dictates salary, but it would be unwise to hire anyone who does not have the same familiarity with the business that you have.

For those of you new to the retail business, this avenue is not recommended, as you have to get your feet wet in the business before you will know who would be as good or better at the job than you.

However, if you can find someone with the experience and want to entrust the management of your new enterprise to this indivdual, that's fine, too. Just keep in mind that you will need to stay involved with the business more at the beginning, at least until a successful business pattern is established and you are confident about leaving the operation of the business in the hands of another.

Recruiting Guidelines

When you hire an employee, there are certain federal requirements that must be met. A certain amount of the employee's wages must be withheld from each paycheck. A separate payroll tax, a percentage of your employee's monthly wage, must be paid. Withholding and payroll taxes must be deposited on a monthly basis, and a state and federal tax form filed quarterly. Check with your bookkeeping consultant or accountant as to proper contract labor procedures in your state.

No matter who you find to help you out, make sure that this person is complimented and properly compensated for good work done. Bonuses are a wonderful incentive to employees and freelancers. And there are many kinds of compensation and recognition—money, lunches, birthday presents, or tickets to concerts—be creative. Find your own way to tell your assistant that you appreciate the help you are getting.

Selection of Candidates

As soon as you feel you are ready, place an ad in the classified section of the local newspaper to recruit potential workers. Also register with your state unemployment office. Students, who will only be available on a seasonal basis but are generally good workers, can be found by contacting placement offices at major colleges and universities. In addition, you can often find qualified people in friends or family members of current employees.

Be prepared to screen quite a few people who are just looking for interim employment or who just don't fit what you're looking for. It takes time to find good workers who are willing to make a commitment to the job. You don't want to waste time training people who are not going to stay.

Have interested applicants fill out a standard employment application (available in any stationery store) and provide you with references from former employers, business associates or, in the case of students, school officials.

Do not just take the information provided at face value; check these references very carefully to find out about the applicant's sense of loyalty, responsibility, and honesty. This can be done by phone or with a letter.

In either case, request answers to specific questions dealing with those factors that are most important to your job requirements; in this case promptness, character, courtesy, dependability, work habits, and loyalty. Be wary of candidates who list friends or relatives as references; they could be trying to conceal unfavorable information.

The Interview Process

Schedule a personal interview to make your own determination based on poise, appearance, level of interest in the job, abilities, and future goals. Write out interview questions ahead of time to help you stay on track.

Set up the interview in a comfortable place to put the candidate at ease.

In addition to finding out about the candidate's capabilities and goals, you will want to use this time to talk about the store,

your expectations, the work standards, and, of course, the pay structure.

On-the-Job Training

When you find someone who seems to have all the qualifications needed, arrange to train him or her on the job. Many owners like to do this themselves, to ensure that their standards are instilled from the beginning and to get a firsthand idea of the new recruit's work habits. If you have a trusted employee, have the new recruit train with your staffer to learn the ropes. (This extra duty should always result in a bonus for the employee doing the training.)

Training may take several days or weeks, depending on the worker's previous experience. Basically, a good training process should involve the following steps:

1. Gain the recruit's confidence by putting him or her at ease.
2. Find out what he or she already knows.
3. Indicate the importance of the specific task being taught.
4. Explain and demonstrate each step patiently.
5. Be sure each step is understood before moving on to the next.
6. Encourage and welcome questions.
7. Have the recruit try to do the task.
8. Correct mistakes gently.
9. Have him or her repeat the steps for you.
10. When you are both comfortable, let the recruit work on his or her own.
11. Review performance periodically.
12. Offer support by letting the employee train others when ready.

Taking time to train properly reduces turnover and accidents, improves the quality of work performed, lowers your cost of labor,

and increases productivity. Every time you have to replace an employee and train a new one, you're losing productivity.

Overtraining the Ambitious

Sometimes it is possible to train people so well they feel they can start their own business in competition with yours. This can occur when an employee realizes you are making money off his or her hard work. There are several ways to stay on top of your employees' activities.

If you're sure an employee is thinking of starting a business, take him or her aside and explain the administrative aspects of the business. Find out if he or she truly realizes that you are providing steady work on a regular basis and if it's really clear how much is involved in running a business.

In rare cases, you are going to lose an employee who feels confident enough to start a business, and there isn't too much you can do about it. In other situations, you can promote the employee, as appropriate. This is feasible only if your cash flow and the employee's ability warrant it.

The Benefits of Happy Employees

Personnel management is a time-consuming job for the business owner. However, paying attention to the needs of your employees and working to gain their trust and maintain their loyalty can do nothing but benefit your business.

The attitude of your employees plays an important part in building and maintaining your store's reputation. If employees are treated fairly and with respect, their job satisfaction will be reflected in the way they do their jobs.

This is something that can truly keep you ahead of the competition; a loyal, efficient, and enthusiastic group of workers is one of your most effective forms of public relations, so never scrimp when it comes to keeping your employees happy. A few important rules of thumb in dealing with employees include the following:

♦ *Never expect an employee to do something you wouldn't do.* This is why training new recruits yourself is such a good idea; it shows them that you are willing and able to step in and do anything that needs to be done.

♦ *Listen to your employees.* Talk to your staff regularly and incorporate any ideas they have whenever feasible. Remember that they have firsthand knowledge about how certain aspects of the job are done and might know if they can be done more efficiently. Suggestions that work for the good of the store should be rewarded with a bonus.

♦ *Take the time to talk about business standards and practices.* Make sure everyone knows exactly what is expected of them. Outline duties and responsibilities on the job and schedule regular reviews to ensure that they are constantly met. If you find it necessary to talk to an employee about work habits, do it in private. And do not criticize; offer constructive ways that performance can be improved.

♦ *Treat each employee as an individual.* When someone seems to be having personal problems that are interfering with his or her work, be willing to allow him or her time off without penalty to take care of the situation. An employee plagued with problems may carry them to work, and this could have a negative effect on productivity. It is much better to get someone to fill in until the regular employee is operating at full efficiency again.

Notes

Key Points

Personal Thoughts

Additional Research

21

ADVERTISING YOUR BUSINESS

More than 150 years ago, Thomas Macaulay, a British historian and statesman, said, "Advertising is to business what steam is to industry. [They provide] the same propelling power."

Few in business would argue with Macaulay's observation—it is as true today as it was when steam was the driving force behind industry. But the question remains, "How do you get the most out of your advertising dollar?" The answer is to

1. Know your customer
2. Target your market
3. Understand the basics of advertising

This chapter provides an overview of various aspects of advertising, including how to use circulation figures to figure your cost per thousand (CPM) and how to create ads that will bring results.

What Is Advertising?

Advertising informs the public about

◆ Who you are

◆ What kind of business you operate

◆ How they can buy your products or services

◆ Why they should come to you

Before you even open the doors of your business, you should start thinking about your advertising program—how much money you can afford to spend, where your dollars will be best spent, and how to structure your campaign.

Decide what kind of results you want. Are you looking for immediate sales or ongoing recognition? What kind of customers are you hoping to attract? Are you emphasizing price, service, workmanship, or something unique? Once you have answered these questions, your decision as to the best type of advertising for the allotted dollars will be easier to make.

There are three basic types of advertising that you will be most interested in during the first few years of your business.

Start-up Advertising

This includes your business cards and stationery, announcement flyers and brochures, and your initial ads in newspapers or other publications. At this point, your main focus will be on telling people where you are located and what you can offer them.

Ongoing Advertising

Once the business is up and running, it will be vital to your success to institute a regular advertising campaign that is well thought out and—this is the key—consistent. Your goal at this point is to attract new clients, obtain repeat business from existing customers, and enhance your reputation.

Appearance Advertising

After you have reached the point where your business is on steady ground and you're showing increased profits every year, you can afford to dabble in "institutional advertising," as it is called in the trade. This is where you pick up the tab to send a dozen kids to the rodeo when it comes to town, or you sponsor a float in the local parade and, in return, get your name listed on the program or on a banner in the parade. This is primarily name recognition only

and, while every little bit is helpful, by the time you can afford it, you probably will be in pretty good shape anyway.

Advertising for Coffee and Tea

Mack Fleming and Bill Hall, who are the only tea growers in the United States, launched an extensive advertising campaign for their new company by sending press kits to newspapers and radio and television stations and by contacting supermarkets, restaurants, hotels, and clubs. Because they were setting a precedent by acting as a food company that controls its product from the field to the supermarket, people were intrigued.

Well-thought-out publicity can definitely enhance your business image and complement your advertising. It will be in your interest to check out your local newspapers to find out what kind of business news they publish. Note whether or not they make use of photographs.

Have photographs taken of your storefront, your products, store renovations, and yourself, and keep some prints and the negatives in your files. Identify the photos for easy future use. Make sure your photographs tell a story. The best kinds of happenings to publicize are listed below.

- ◆ Book you have written
- ◆ Business anniversary
- ◆ Business awards
- ◆ Charitable activities
- ◆ College scholarships you offer
- ◆ Contest sponsorships
- ◆ Grand opening
- ◆ Quotes from industry sources
- ◆ Trade award
- ◆ Unusual services/products

Brochures

A brochure is a wonderful way to promote the products you have
to sell. You can list the coffees and teas that you specialize in and
include a few paragraphs with regard to the differences in each of
the blends you carry. People enjoy getting information in general
and coffee and tea gourmets, especially, care enough about the
beverages to want to know more about them.

In your brochure you also have the opportunity to mention
your special services, such as mail-order coffee packs and gift
baskets, if you make these. Look at the brochures that you see in
other gourmet food and gift stores and note what works. This will
help you design your own.

Other types of advertising include the following:

♦ **Television.** This is quite expensive for a small business
owner, but if you can afford it, do it. The effectiveness of
the exposure is worth the cost.

♦ **Local Radio.** This has always been an effective way to get
a message to the public. The cost is based on the amount
of time used and the time of day. Trade-off plans (also
known as "trade-outs") are sometimes used whereby the
ad cost is paid in barter or through a tie-in promotion, such
as a contest, sponsored by the station with prizes donated
by your store.

♦ **Buses, Transit Cars.** If your city has public transporta-
tion, this can be a good method of advertising. The public
gets the message every time they ride!

♦ **Reviews.** A food editor or a restaurant critic can be called
to come in, make an honest evaluation of the coffees and
teas, and write an article based on his or her impressions.

Advertising is a continuing effort. Someone once said, "Advertis-
ing costs money, but it is not expensive." This is largely true,
especially when it is effective advertising. While you eventually
will be able to rely on word-of-mouth advertising, you should
nevertheless plan to set aside a certain sum each month for a

regular advertising program; between 1 and 5 percent of projected gross sales is a good starting point.

Direct Mail

Keep in mind that you promote your business through publicity, advertising, direct-mail, or personal contact. Direct mail is often the best, because a prospective customer can read your "approach" letter at his or her convenience. Sales letters are the least expensive direct-mail pieces to produce, but they are far from the easiest to write. You may want to hire a copywriter to handle that task for best results. In creating your letter, bear in mind the following points:

1. The letter introduces your products and services to potential buyers.
2. It must have a dynamic lead-in sentence.
3. It must present crisp and clear information about the benefits of dealing with you.
4. A feeling of personal communication, that it was written just for the person reading it, must be conveyed.

Think of what it takes to draw you into a direct-mail piece and keep you interested until the last word: a thought-provoking phrase in big, bold letters at the top of the pages; a simple but thorough explanation of why the letter was sent and how reading it would be of benefit to you; a direct request to take action; a few testimonials from the service's users; the signature of the company's president. Consider all of these factors when preparing a direct-mail sales letter.

On the following page is a sample letter that you might want to examine and change for your own use.

Sample Direct-Mail Letter

THE COFFEE CAFE
123 Main Street, Anytown, USA
555-1212

Dear Friend:

I can't function in the morning until I've had a cup of steaming hot, custom-blended Hawaiian Kona Coffee with fresh cream. That's why I don't open The Coffee Cafe until 7:30 A.M.—I spend the first hour I'm here creating exciting new blends for the day, baking a batch of bran muffins for the morning crowd, and sitting back with my first cup of the day.

You may be like me—looking forward to that rich flavor of a freshly brewed cup of java in the morning. If so, I'd like to invite you to become a member of The Coffee Club. Membership entitles you to incredible benefits that any coffee lover would appreciate, including a 20 percent discount on each pound of *whole bean* coffee; a 15 percent discount on each pound of *ground* coffee; a 30 percent discount on grinders, brewers, and coffee pots; our quarterly newsletter; and an open invitation to join me and your fellow club members every morning between 7 and 7:30 A.M. for a special cup to start the day off right.

Membership in the club is only $10 per year and can save you hundreds of dollars on your personal and gift gourmet coffee purchases! Call me today for further information or to sign up as a charter Coffee Club member. I look forward to adding the pleasure of truly fine coffees and teas from around the world to your day!

Sincerely,
Joe Miller, Owner

Also consider having an inexpensive flyer made up with your name, address, and an outline of your coffee selections included. Hire someone to put them on car windshields in the parking lots of major shopping malls or on doors of homes and apartments in your area. As a way to "read" the response, include a coupon on the flyer or talk about something unique, such as "Ask about our 3:00 P.M. pick-me-up coffee and Danish special."

Some stores use specialty promotional items, such as telephone numbers printed on calendars, pens, key chains, and jumbo paper clips. This is one way to remind clients who you are and where you are located. However, the best way to promote your store is through your consistent, topnotch service and high-quality products.

Yellow Pages

You might want to consider listing your business in the Yellow Pages under the appropriate classification. Check with the phone company and find out when their next directory is coming out and what the deadline is for a Yellow Pages ad. Buy the biggest ad that you think you can afford. Experience indicates that these are worth the investment.

Classified Advertising

Don't underestimate the power of classified ads. Many major corporations use the classifieds even though they have sizable budgets available for display advertising. There are several reasons for this.

1. The classifieds are an extremely reliable testing ground for new products, services, and ideas. Although it is true that people who typically read the classifieds are a different group from those who scan display ads, they are considered to be responsive and, therefore, can give you a very good idea of whether or not you have placed your ad in the appropriate publication.

2. A short, well-written classified placed in the right publication and under the proper category can be a low-cost method of advertising that guarantees solid returns.

3. If your company is trying to establish a mailing list, a classified ad that features an "Inquiry" statement, such as "Send name and address for free details (or a brochure)," is a good way to build up a file of qualified buyers' names. And they can be considered qualified buyers because it takes time, energy, and the cost of a postage stamp for them to get your free information. By writing to you they have stated their interest.

4. Classified ads are inexpensive, ranging from $0.50 cents to $15.00 per word, depending on the publication. With careful planning, you should be able to get broad-based coverage without putting a dent in your operating capital.

Here's an example:

FREE!! 1 pound of 100% Colombian Coffee Beans with every 10 pounds you buy. Enjoy a complimentary cup of freshly brewed Mocha Java this week, on the house. Just ask for Joe, The Coffee Cafe, 123 Main Street.

Display Advertising

Don't dismiss the possibility of running small display ads on a seasonal basis, even though they are more expensive than classified advertising. Consider offering an incentive, such as a 10 percent discount if customers bring in the ad. This will help you track responses and plan future display ad campaigns.

As with all types of advertising, it is important to define your market when getting ready to place a display ad. Your main goals are to select a publication that will reach the audience you want and then to create a specific, attention-getting ad that appeals to that target group.

Anytown's

Largest Selection of Gourmet Beverages From Around the World

By the cup or by the pound

Large Selection of Decaffeinated Coffees

Over 100 Teas

Unique Gifts

Full Line of Beverage Accessories

THE COFFEE CAFE

123 Main Street
555-1212

To Write or Not to Write

Writing display ad copy is not for the inexperienced. Although it is possible to learn how to put ideas and words together that will pull the results you desire, it is recommended that you hire a copywriter if you have any qualms about producing an ad.

However, if you are confident that you can develop your own ad, remember that it must build identity through the use of carefully planned words and design. Glance through newspapers and magazines to find ads that catch your eye. Most likely, the ones that stand out have a memorable logo, effective use of space, or an unusual headline that leads you into the body of the ad.

These are ads you should refer to as samples when designing your own.

When planning your ad, keep the following elements in mind:

1. *Visibility.* Your ad may be surrounded by many others, so design it so that it will immediately attract the reader's attention.

2. *Boldness.* Use large art and/or a bold headline as a focal point.

3. *Simplicity.* Don't overwhelm the reader with too many fine details. The ad's main point is lost and so is the reader's attention. This is particularly true in a small ad.

4. *White space.* Just because you have, say, a four-by-six-inch space to work with, it isn't necessary to fill it up. In fact, white space is a necessary component in assuring your ad will be read.

5. *Use simple-to-read typefaces.* Two that are popular are Times Roman and Bookman, which are known as serif typefaces because of the tiny strokes at the tops and bottoms of the letters. Sans-serif (without strokes) type-faces, such as Helvetica, are fine for ads containing only a few words, but are difficult for the eye to follow when there is a lot of text.

 Also, be sure that the type is large enough; generally nothing smaller than 10-point type should be used. If type is too small to read, it will simply be ignored.

Design and Typesetting

It isn't necessary to be a great artist to create an ad, especially these days with the availability of a variety of impressive graphic software programs. In addition, you can buy cut-out-and-transfer (press-on) letters in different typefaces, symbols, borders, and designs through graphic-art-supply companies. Most word processing programs also contain computer graphics that can dress up your ad at a low cost.

If you feel uncomfortable about laying out your ad so it has eye appeal, consider hiring an art student to handle the job for a prearranged fee or as a school assignment. (Talk with the head of the art department to see if there is a work/study for credit program.) Just be sure to review the student's work prior to making a commitment.

Also, check with the advertising department of the newspaper or magazine you are planning to advertise in. There may be graphic artists or designers on staff who will work on the layout for you. In fact, there are still newspapers in the country that offer full service, from ad concept to design work, at a minimal charge to their advertising clients.

> *Advertising is the greatest art form of the twentieth century.*
>
> *Marshall McLuhan*

Publications work on tight deadlines, so be sure you start the process early enough to get a proof copy of your ad back in time to make any corrections. You can imagine the anguish in seeing your ad appear with the wrong address. Although the publication would probably do a "make-good" for you and run a corrected ad at no charge, the damage has already been done. The final responsibility for the ad rests with you, so, as with any other aspect of your business, plan ahead.

Tracking Ad Response

Some customers will tell you that they saw your ad and might even let you know what they liked or disliked about it if you ask them. Even so, you should develop other methods for determining if your advertising is working.

One very simple way is to include a coupon for something in the ad and then count the number of coupons you get within a certain test period after the ad runs. There is one major problem with this, however. Even the most well-intentioned people often cut coupons, file them away in a "safe" place, and totally forget about them.

So, although you will be able to gauge response to some degree, be aware that many of the people who come into or call your business have probably seen the coupon but simply mislaid it or are not the kind of folks who use them.

Predicting Ad Response

There is a standardized formula in advertising that provides a barometer for predicting how much response can be expected from either a display ad or a classified ad. The formula states that you will see one-half of the total responses from an ad within a certain period of time after receiving your first inquiry or order.

For an ad run in a daily newspaper, the period of time is three days; for a weekly newspaper or magazine, it is six days; within fifteen days for a monthly publication, and within twenty-five days when running in a bimonthly. Although there will be exceptions to this rule, it does give you a base from which to track response.

Determining Your Cost Per Thousand

The cost per thousand (CPM) equation helps you develop a cost-effective campaign. Basically, it tells you what your ad cost per thousand readers will be. Most publications will provide a CPM comparison upon request (some include it in their media kits), but you can figure it out for yourself with just a few facts from publications that you are exploring as advertising vehicles.

For convenience sake, CPM equations are typically based on the rate of a full-page, black-and-white ad. You simply divide the full-page rate by the thousands of circulation. It is important that you get the circulation, *not* the readership, because magazines and newspapers typically claim that their readership is 2 to 50 percent higher because of "pass-along" of the publication to friends or relatives.

For example, if a certain newspaper is charging $2,000 for a full-page ad, and it claims the true circulation is 200,000, you will be paying $10 per 1,000 readers for your ad space. Another specialized publication's full-page rate may be $1,000 with a circulation of 50,000. The cost per 1,000 readers will be much higher—$20 per 1,000, but it might be worth it if, for example, you have a unique product or service that is geared to an exclusive market.

Benefits of Paid Circulation

It is also important to know that publications with a paid circulation generally have a readership that is more inclined to respond to advertising. This is because of the simple fact that they are a captive audience who have taken the time to *order* the publication. This is especially valuable if you have a product or service that you're planning to market through mail-order channels.

You can find circulation, readership demographics, advertising rates, and other important information about a number of publications (especially those with national distribution) through *Standard Rate and Data Service (SRDS)*, available through the research desk of your local library.

The Advertising Research Foundation and the Association of Business Publishers conducted a study to determine the impact of advertising on the sale of products. Several different products were used for the study, and each was advertised for a twelve-month period in an appropriate publication. The results were interesting, but not surprising to anyone who has ever utilized a solid advertising campaign in promoting a business.

♦ More advertising meant more sales.

♦ Determining results from an ad campaign generally took four to six months. One or two insertions do not indicate viable results.

♦ Color in advertisements dramatically increased response and sales.

♦ A well-developed ad campaign keeps on working for a year and sometimes even longer in publications with high "keep" appeal.

Knowledge and belief in your product, faith in yourself, and respect for your customers are the keys to successfully building your future. As you go about starting and establishing your business, remember the word "profit." This alone should give you the necessary motivation to get out there and confidently tell the world what you have to offer.

Notes

Key Points

Personal Thoughts

Additional Research

22

PROMOTION AND PUBLIC RELATIONS

Informing the public about your business through the use of business cards, brochures, mailing pieces, catalogs, and specialty items (calendars, pens, and T-shirts imprinted with your company name) is a form of advertising that is known as promotion; what you do over and above your paid advertising and promotion that help build your image and keep your business in the public consciousness are referred to as public relations. There is a fine line between the two, and they can make or break your business if not addressed.

There are many clever ways to extend the effectiveness of your advertising and promotional dollars, as illustrated in the following examples.

A Little Creativity Goes a Long Way

The owner of a pet grooming business leaves a business card and a brochure featuring a 20 percent discount coupon everywhere he goes. When he is running errands, he always takes a handful of brochures with him to hand to store clerks, gas station attendants, and even food servers he runs into during the day.

If he sees a car with a dog in it, or one with a bumper sticker announcing the owner's affection for a pet, he slips a brochure

219

under the windshield wiper. Does it work? Absolutely. He claims that 45 percent of his new business comes from the recipients of his handouts, and the majority of them become regular customers.

Another business owner who operates a small walking tour service in her beachside town sends out a one-page quarterly newsletter featuring historical and other facts about the area to everyone who has ever taken the tour. She includes a $10 coupon that can be redeemed by former clients or their guests.

She states that many of her clients are local residents who send their out-of-town visitors on her walking tour, simply because she makes sure they are aware that she is still in business and generates enthusiasm through the newsletter.

Advertisements contain the only truths to be relied on in a newspaper.

Thomas Jefferson

Charity Tie-ins

Other ideas you might consider include promoting humanitarian outreach. For example, for every ten products a customer buys or every tenth time a customer uses your service, you donate a predetermined amount of money to a local charity. You can easily keep track of customers' purchases by punching a hole at the edge of a card designed by you or your printer strictly for that purpose.

When you are ready to present the check to the particular charity, make sure the chairperson of the organization is there to accept it and be sure to contact the local press and invite them to the "event." In most cases, you'll receive free coverage.

Reminder Cards

Sending regular "thinking-of-you" cards to your past and present clientele is especially effective with service businesses. Again, it assures the customer that you are friendly, reliable, and conscientious—real pluses for your credibility.

Promote Your Grand Opening

The grand opening is a great promotion tool if it is handled properly. One of the most frequent and costly mistakes made by inexperienced operators is to plan for a gala grand opening celebration *before* they are ready.

The urge to do this is based on the belief that the celebration will result in a lot of good free publicity. What usually happens is that the crowds are tremendous, the service is poor, and the high expectations of customers are dashed by the failure of the new operation to cope with the volume of business.

How do you do it successfully? Open quietly, without fanfare, *especially if this is your first venture.* Then after a few weeks, when you have all the kinks worked out of your system, announce a grand opening. At that time, go all out for publicity. Try to choose a prominent group to invite to your opening from the market you are trying to serve. Are there any noted chefs in the community? Invite them, as they are often coffee connoisseurs. Does the Mayor like coffee and, if so, what kind? You can find out by calling his or her assistant. Then send a special invitation indicating that you will have the Mayor's special blend available at the grand opening. (Be sure to invite the assistant, too, as he or she can be a powerful source of word-of-mouth advertising.)

Other promotional ideas can include the following:

1. Two-for-one offers, i.e., buy one cup, get a second for a friend at no additional charge.

2. Printed matter with the name of your coffee shop to give away to customers, such as matchbooks, mugs, or coffee history cards designed to be kept by the customer.

3. A very successful promotion technique is to have a certain night during the week designated "ethnic night." On those nights, you could feature certain coffees and teas from various ethnic areas. If possible, try to tie it in with a local bakery that may be able to provide sweets and breads from the same country or region.

Networking

Check out local business-owner groups and the chamber of commerce in your area. Pay your membership dues and join as many groups as possible. Membership in these organizations offers you the opportunity to meet people who might use your products and services and will, at bare minimum, tell others about you once they know you, feel comfortable with you, and understand what you are offering. You will also be given a listing in the group's directory, generally under your specific category.

And, as is so often the case, members are prone to supporting others involved with their group. It is quite possible that members with compatible businesses will put a stack of your cards or brochures on their counter or in their referral file. In return, you may be able to promote their businesses to your customers. This is the true meaning of networking.

The directory also gives you bona-fide access to an instant mailing list, which you can use to send out promotional flyers or brochures. Participating in the organization's events and the willingness, for example, to speak at functions about your area of specialization will let people know that you are community-minded. This involvement will work to project a positive image for you and your business.

Customer Service

One of the most overlooked areas in promoting business is the way we create an impression when dealing with the public on a one-to-one basis. The ageless philosophy that the customer is always right and comes first has, it seems, gotten lost in the shuffle in these days of fast-paced transactions.

As a small business owner, you can be sure that customers will flock to you if you make them feel important. It is as easy as greeting them warmly (instruct your employees, if you have them, to do the same), maintaining a courteous attitude, and inviting them to come back soon—even if they have not spent any money.

Remember that word travels fast, and if you create an atmosphere that makes all your customers feel special, you can be sure they will tell their friends and neighbors.

Free Publicity

Free publicity also comes under the category of public relations. This includes newspaper, magazine, radio, and television interviews that feature interesting or unusual facts about you and your business.

As any newspaper reporter or talk-show host will tell you, everybody has a story to tell. The key is to get the media to zero in on yours and make it available to the public. If you have trouble deciding on a unique angle for your story, invite several friends over for a brainstorming session to create a newsworthy item.

Use your imagination to explore the creative possibilities of your business venture. And then look at your personal story. Perhaps you have completely switched fields with your new business: There's a story with a strong slant toward risk-taking. Maybe you have successfully turned a hobby into a business venture, created a unique product, or developed a new twist on an old theme. These all qualify as human-interest stories. Newspaper and magazine editors like them almost as much as their readers enjoy them.

The Press Release

Once you have your story angle, call the managing editor or the feature editor of your local newspaper. Give him or her a brief description of who you are and what business you are in. Say that you will be sending out a press release (see following sample) and would be happy to arrange an interview at a convenient time.

Sample Press Release

FOR IMMEDIATE RELEASE
MORE INFORMATION: 555-1212

The Coffee Cafe Offers International Treasures

Joe and Rose Miller, owners of The Coffee Cafe, are pleased to announce the opening of their new coffee and tea store, located at 123 Main Street. A Grand Opening celebration will be held next Friday.

"We are planning an exciting opening," said Miller, "featuring mimes, free coffee and tea, baked treats from Sal's Country Bakery, and lots of prizes, including a $500 home espresso machine."

Miller is a former corporate personnel director who became interested in coffees of the world after trying his first custom blend on a trip to South America. "I was overwhelmed," he said, "after my first sip of a Peruvian/Colombian blend the waiter in the hotel brewed up. It made the commercial brands I had been drinking for so long seem like tap water."

The Millers now consider themselves "coffeeologists" as they have devoted the past ten years to studying international coffees and teas (teas are Rose Miller's specialty based on her English heritage), perfecting the blends that have won them six awards from brewer's associations, and preparing to open their coffee and tea store.

"The store is our dream come true," Joe Miller said, "and our goal is to pass our excitement along to the customers." For further information on the grand opening of The Coffee Cafe, call Joe at 555-1212.

Invite journalists to visit your place of business to get a firsthand look at what you are doing and how you do it. Do the same with program directors or talk-show hosts at local and regional television and radio stations.

If, on the first attempt, your presentation fails to result in an article, follow up in three to six months, possibly with a new story angle. In the meantime, however, send out regular press releases announcing new developments with your business—the grand opening, special events, details on your business philosophy, extra services, and features.

Even if they don't run a full story on you, there is a good possibility you could be included in a feature story about local entrepreneurs or people in related businesses.

Previously, we discussed the advantages of opening quietly for a week or two before your grand opening, just to get the kinks out of the operation. During that period, you might invite representatives of various news media to have coffee with you, individually, so you can tell them about your hopes and plans for the new operation.

Suggest a taped interview, or a feature story. At the same time, if you can afford it, arrange for some concentrated advertising during the week before the opening and hopefully at the same time the article on your coffee and tea store is published. Your adjunct advertising campaign might go like this:

Saturday. Watch for the opening of The Coffee Cafe! Look for Monday's announcement.

Monday. Keep Thursday night open! Gala opening of The Coffee Cafe. Fresh roasted coffee from around the world. 123 Main Street.

Tuesday. The Coffee Cafe. Grand Opening—Thursday, January 24, 5:30 P.M. to 9:30 P.M. Gala buffet featuring European coffee and tea blends with baked treats from Sal's Country Bakery. 123 Main Street. 555-1212

Wednesday. Same ad as Tuesday, except print entire listing of coffee and tea blends available.

Thursday. Repeat Wednesday's ad.

This series, which started with "teaser ads" and ended with full information about the coffee store and your opening festivities,

could be read on the local radio station as well. The station could also read part of the listing of coffee blends and feature an interview with you to share your expertise about coffee and tea. Your newspaper ads could be enlivened with a sketch of a group sharing a coffee break, or with a photo of your new place looking cozy and inviting.

An Effective Ad Campaign

All of the methods outlined earlier, along with any others you may develop, will keep your name in the consumer's mind. The effectiveness of an advertising and/or publicity campaign can be measured by conducting a simple marketing survey with new customers. Simply make it standard practice to ask each customer where he or she heard about you and your business.

Keep a tally of the responses in a notebook. A periodic review will give you hints of where to allocate future advertising funds. If, for example, the majority of your customers are being drawn from your display ad in the local newspaper, keep it going on a steady basis. The same principle applies to classified, direct-mail, or word-of-mouth advertising.

The justification for investing in advertising and promotion is time. If you attempted to contact all of the people who read ads and press-release material in newspapers, or those who listen to talk shows, you would never have time to conduct your business.

The advertisement is one of the most interesting and difficult of modern literary forms.

Aldous Huxley

Time is money! As a small business owner, you will want to devote as much attention as possible to the production end of your venture and let your advertising and promotion work bring in the customers.

Getting Started

At this point, you're ready to go out there and give your coffee and tea business a try. If you've done your homework, by now you should know:

♦ What products and services you plan to provide

♦ Your prices

♦ Estimated start-up and equipment costs

♦ Resources for equipment

♦ How to set up the space

♦ How to get started with advertising and promotion

It is absolutely crucial that you approach your start-up with good planning and organization. Take each step knowing what the previous step was and which step comes next. Likewise, know what you want to happen and what could happen; plan for problems, and have possible solutions ready.

This book just scratches the surface of the knowledge that you will want—and need—to acquire in the course of starting your business and making it grow. There are many resources to help you learn and stay abreast of the industry's latest news. Some of these resources follow in the Appendix.

For some time, until your coffee and tea store is very well established, you will be the manager as well as the chief brewer and bottle washer. You want to be the very best manager you can be. Review the following checklist of traits and skills you will want to develop to ensure success.

Guidelines for a Successful Operation

A manager should

1. Relate well with all people

2. Portray an air of confidence, not one of crisis or panic

3. Be able to keep composure under all circumstances

4. Seek methods to promote productivity among employees

5. Not be merely a "nine-to-five" person; must be willing to give as much time as necessary for success—in other words, be dedicated to the business

6. Work the hours the business calls for, even holidays

7. Always be aware of sales and figures, and be aware of how the store is doing now, last week, last month, and how you want it to be doing tomorrow, next week, next month

8. Talk with customers

9. Lead your staff by setting a good example

10. Always be ready to improve service

11. Be a good listener and keep an open mind

12. Use input from customers to improve the business

13. Always be prepared to write down customer information on a pad carried in pocket or on clipboard

Good luck to you in your new endeavor.

Notes

Key Points

Personal Thoughts

Additional Research

Appendix: Resources

Trade Associations

Green Coffee Association of New York
116 John Street, Suite 1705
New York, NY 10038
(212) 619-5095

Represents United States and Canadian brokers, importers, and roasters. Arbitrates green coffee contracts and publishes statistics and bulletins.

National Coffee Association
110 Wall Street
New York, NY 10005
(212) 344-5596

Represents the major United States' coffee companies represented in the food service and retail markets. Publishes technical studies, surveys, and position papers. Primarily concerned with coffee as a commodity.

National Coffee Service Association

4000 Williamsburg Square
Fairfax, VA 22032
800-683-NCSA

Represents the coffee service industry that provides coffee, tea, refreshments, and service equipment to restaurants, hotels, gourmet coffee shops, hospitals, and corporate facilities. Promotes and advances mutual interests of members by providing educational materials, industry research, and industry-specific publications.

Pacific Coast Coffee Association

Ferry Building, Suite 3037
San Francisco, CA 94111
(415) 986-0267

Represents distributors, importers, and processors of green or roasted coffee, primarily on the United States' West Coast. Main activities include the publication and arbitration of green coffee contracts.

Southern Coffee Association

c/o InterContinental de Cafe International
160 Water Street, 13th Floor
New York, NY 10038
(212) 514-6600

Meets twice a year to share information among coffee professionals.

Specialty Coffee Association of America

One World Trader Center, Suite 800
Long Beach, CA 90831
(310) 983-8090

Nonprofit trade association comprised of 1,500+ international members (distributors, exporters, growers, importers, retailers, roasters, and representatives of allied industries) from coffee-consuming and coffee-producing nations. Objectives include the promotion of high-quality coffee beverages,

education about specialty coffee, and provision of a forum for discussion and unification among members of the specialty coffee industry.

Associations and Agencies

Small Business Associations & Government Agencies

American Marketing Association
250 South Wacker Drive, Suite 200
Chicago, IL 60606-5819

Marketing publications available to nonmembers.

Bureau of the Census
Washington, DC 20233

Statistical data.

Chambers of Commerce
Consult Yellow Pages. Services vary among chambers but may include advice, support, workshops or seminars, local business news, and other services specifically for small businesses.

Council of Better Business Bureaus
4200 Wilson Boulevard, Suite 800
Arlington, VA 22203

Ask for a listing of their "Booklets on Wise Buying."

Dun & Bradstreet
299 Park Avenue
New York, NY 10171

Send for the booklet "This is Dun & Bradstreet," an overview of publications and services.

Internal Revenue Service (IRS)
1-800-829-1040 General tax assistance
1-800-829-3676 Forms and publications

See the government section of your phone book for address and phone of the local office. The IRS publishes a variety of books and brochures that are useful to owners of small businesses. Each year the IRS publishes *Tax Guide for Small Business*.

International Franchise Association
1350 New York Avenue NW
Washington, DC 20005

Regulation and information on franchises.

Local Planning Agencies
See the government section of your phone book for address and phone of the local office. Source of information on zoning and future development locations.

Minority Business Development Agency
Office of Public Affairs
Department of Commerce
Washington, DC 20230

National Association for the Self-Employed
PO Box 612067
DFW Airport, TX 75261-2067

National Association of Women Business Owners
600 South Federal Street
Chicago, IL 60605

National Federation of Independent Business
150 West 20th Avenue
San Mateo, CA 94403

National Insurance Consumer Organization
PO Box 3243
Merrifield, VA 22116-3243

Send $3 and self-addressed, stamped #10 envelope for book-let, "Buyer's Guide to Insurance."

National Minority Business Council
235 East 42nd Street
New York, NY 10017

Bimonthly newsletter for small and minority businesses.

National Small Business United
1155 15th Street NW, Suite 710
Washington, DC 20005

Send for information on federal legislation.

Occupational Safety & Health Administration (OSHA)
Department of Labor
200 Constitution Avenue NW
Washington, DC 20210

Office of Information and Public Affairs
U.S. Department of Labor
200 Constitution Avenue NW
Washington, DC 20210

Request publications list regarding employment regulations.

Service Corps of Retired Executives (SCORE)
See Small Business Administration.

Small Business Administration
1441 L Street NW
Washington, DC 20416

The SBA is the primary source of federal governmental assistance for small businesses. In addition to providing financing for many small businesses, it provides a number of services for free or at a small cost. SBA regional offices routinely conduct free or low-cost small business workshops. The SBA also offers many helpful publications as well as the following hotlines:

SBA Answer Desk

Recorded information on a variety of topics

1-800-827-5722

SBA Online

A national electronic bulletin board service that provides an overview of SBA information.

1-800-859-INFO if you use a 2400 baud modem or 800-697-INFO if you use a 9600 baud modem

See also following entry.

Service Corps of Retired Executives (SCORE)

PO Box 15434
Fort Worth, TX 76119

Retired business executives offer their expertise and services to entrepreneurs through SBA's SCORE, which charges for out-of-pocket expenses but not for the consultant's time.

Telephone Company

Consult your local phone company. Most offer a variety of services and publications for small businesses.

AT&T 800-Number Information

Call 1-800-272-0400.

Books

General Reference Books & Periodicals

Note: Many of the following resources are available through the Research Desk at your local library.

A Consumer's Guide to Telephone Service

Consumer Information Center, Pueblo, CO 81009

Bacon's Publicity Checker
A comprehensive listing of every major newspaper in the United States and Canada.

Directory of Publications and Broadcast Media
Published annually by Gale Research Company, Detroit, MI.

The Dun & Bradstreet Middle Market Directory and The Dun & Bradstreet Million Dollar Directory
The Dun & Bradstreet Corporation, New York, NY.

Encyclopedia of Associations
Published annually by Gale Research Company, Detroit, MI.

Encyclopedia of Business Information Sources
Published periodically by Gale Research Company, Detroit, MI.

Financial Studies of the Small Business
Financial Research Associates
PO Box 7708
Winter Haven, FL 33883

National Five-Digit ZIP Directory (2 volumes)
U.S. Postal Service
6060 Primacy Parkway, Suite 101
Memphis, TN 38188-0001

Also available by phone (1-800-238-3150) or through your local post office.

Poor's Register of Corporations, Directors, and Executives
New York, NY: Standard & Poor's Corporation.

Tax Guide for Small Businesses
Internal Revenue Service publication. Published annually. Available from local IRS office.

Thomas Register of American Manufacturers
Thomas Publishing Company, New York, NY.

Ulrich's International Periodicals Directory
Comprehensive listing of major magazines and newspapers.

Where to Find Business Information
by David Brownstone and Gorton Carruth
New York: John Wiley & Sons, 1982.

Useful Books for the Coffee and Tea Store Business

(Note: The Specialty Coffee Association of America has a product catalog that offers a variety of books, brochures, videos, and promotional items. See SCAA listing that follows for sample titles.)

Coffee: A Guide to Buying, Brewing, and Enjoying
by Kenneth Davids
Santa Rosa, CA: 101 Productions (Cole Group imprint), 1991

Coffee, Beans to Brew: A Cookbook for Coffee Drinkers
by Mary Ward
Hollywood, FL: Lifetime, 1992

The Complete Coffee Book
by Sara Perry, et al.
San Francisco: Chronicle Books, 1991

The Perfect Cup: A Coffee Lover's Guide to Buying, Brewing, and Tasting
by Timothy J. Castle
Reading, MA: Addison-Wesley Publishing, 1991

The Republic of Tea
by Mel and Patricia Ziegler
New York: Doubleday, 1992

Specialty Coffee Association of America Titles:

Avenues for Growth

The Basics of Cupping Coffee

The Coffee Cupper's Handbook

Green Coffee Contract & Arbitration Handbook

Retailer Executive Compensation Report

Magazines

General Business

Business Marketing
Crain Communications, Inc.
740 N. Rush Street
Chicago, IL 60611

Published monthly.

Entrepreneur Magazine
2392 Morse Avenue
Irvine, CA 92714

Published monthly.

Sales & Marketing Management
Bill Communications, Inc.
355 Park Avenue South
New York, NY 10010

Published monthly except in June, August, and October, when it is published twice a month.

Success, The Magazine for Today's Entrepreneurial Mind

Lang Communications
230 Park Avenue
New York, NY 10169

Published monthly with the exception of bimonthly issues for January/February and July/August.

Working Woman

Lang Communications
230 Park Avenue
7th Floor
New York, NY 10169

Published monthly.

Trade

In Good Taste

Specialty Coffee Association of America
One World Trader Center, Suite 800
Long Beach, CA 90831

Newsletter for SCAA members. Published ten times per year. Circulation 1,600.

Tea & Coffee Trade Journal

130 West 42nd Street
New York, NY 10036-7802

Published monthly. Features industry news including events and personnel, new products. Circulation 7,500.

World Coffee & Tea

1801 Rockville Pike, Suite 330
Rockville, MD 20852

Published monthly. The premier trade publication serving the coffee and tea industry. Articles directed at specialty coffee

and gourmet tea retailing, office coffee service, roasters and brand packagers, exporters/importers, equipment manufacturers, etc. Circulation 8,000.

Suppliers

Coffee and Tea

Coffee Bean, The
10020-D San Vicente Boulevard
Los Angeles, CA 90049

Crabtree & Evelyn, Ltd.
Box 167
Woodstock Hill, CT 06284

East India Tea & Coffee Co., Ltd.
1481 3rd Street
San Francisco, CA 94107

First Colony Coffee & Tea Company, Inc.
204-222 West 22 Street
PO Box 11005
Norfolk, VA 23517

Fortnum & Mason, Ltd.
Piccadilly, London
England W1A 1ER

Fox Mountain Farm
1600 Chippewa Street
New Orleans, LA 70130

Grace Tea Company, Ltd.
80 Fifth Avenue
New York, NY 10003

Graffeo Coffee House
733 Columbus Avenue
San Francisco, CA 94133

McNulty's Tea & Coffee Company
109 Christopher Street
New York, NY 10014

Northwestern Coffee Mills
217 North Broadway
Milwaukee, WI 53202

O.H. Clapp & Company, Inc.
42 Riverside Avenue
Westport, CT 06880

Pannikin
7467 Girard Avenue
La Jolla, CA 92037

Paprikas Weiss Importer
1546 Second Avenue
New York, NY 10028

Peet's Coffee, Tea & Spices Company
2124 Vine Street
Berkeley, CA 94708

Starbuck's
By Mail
800-445-3428

Espresso Brewers

Thomas Cara
517 Pacific Avenue
San Francisco, CA 94133

Cold Water Brewers

Toddy Products, Inc.
1206 Brooks Street
Houston, TX 77009

Filter Brewers

Melitta, Inc.
1401 Berlin Road
Cherry Hill, NJ 08003

Paprikas Weiss Importer
1546 Second Avenue
New York, NY 10028

General Beverage Accessories

Cross Imports
210 Hanover Street
Boston, MA. 02113

Grinders

Nelson & Sons, Inc.
PO Box 1296
Salt Lake City, UT 84110

INDEX

A

Accountants, 95

Acidity, definition of, 182

Ad copy
 design of, 213
 writing of, 212

Adding machine, 37

Advertising
 of business, 203-216
 campaign, 226-228
 classified, 209-210
 for coffee/tea store, 205-216
 cost of, 214-215
 display, 210-211
 of grand opening, 68, 221
 and media kits, 74-75
 response to, 214
 types of, 203-205

African family, of coffee, 181

Agencies, for small business, 233-236

Appearance advertising, 204-205

Apple Macintosh, compared to IBM Windows, 51-52

Applicants, for employment, 197

Application, business capabilities, 19-23

Arabian family, of coffee, 181

Arabica coffee beans, 178

Aroma, of coffee, 178, 182

Assets, on balance sheet, 145

Associations, for small business, 233-236

Atmosphere, of coffee/tea store, 28, 84

Attorneys, as professional backup, 94

Auctions, for buying equipment, 61-62

Audit, by IRS, 100

Automatic filter drip, brewing method, 186-187

B

Balance sheet, 120-121, 145-146

Bank statement, reconciliation of, 146-147

BCA, 19-23

Beans, coffee, 178

Small Business Administration
(SBA), 124, 128-129
Software, for computer, 49
Sole proprietorship, 103
Souchong, 190
Specialty Coffee Association of
America, 25
Specialty coffee store, *see* Coffee/tea store
Standard Rate and Data Service
(SRDS), 215
Start-up advertising, 204
Start-up costs
for cart vendor, 43
for coffee/tea store, 26, 44
for small business, 35-36
worksheet for, 46
State income taxes, 101
Stationery, for business, 112
Stockholders, 104
Storage capacity, of computer, 50
Subchapter S corporation,
104-105
Success tips, for coffee/tea store,
156
Suppliers
of coffee and tea, 241-242
of equipment, 242-243
Supplies
for cart vendor, 30, 43
for coffee/tea store, 41
Surge suppressors, 55

T

Tasting, coffee, 180-182
Taxes
deductions for, 100-101

record keeping for, 99-100
state and local, 101
Taxpayer identification number,
102
Tea
auctions for, 191
blending of, 191
gift baskets of, 158-159
herbal, 192
leaves of, 190
packaging of, 192
price of, 157
processing of, 189-190
Tea bags, 192
Tea store, *see* Coffee/tea store
Telephones, 38
Television, for advertising, 206
Time, managing, 171-174
Trade, books for, 240
Traffic patterns, 87
Training, on-the-job, 198
True lease agreement, 63-64
Turnkey operation, 145
Typeface
for ad copy, 212
for logo, 110-111, 113
Typesetting
of ad copy, 213
fees for, 111-112

U

Used equipment, 59-62

V

Vacuum, brewing method, 186
Venture capitalists, 129-130
Visibility, high, 67-69

Other Titles in the Start Your Own Business Series:

Start Your Own Desktop Publishing Business 0-89384-245-1
Start Your Own Gift Basket Business 0-89384-261-3
Start Your Own Import/Export Business 0-89384-248-6
Start Your Own: 101 Extra Income Ideas 0-89384-265-6
Start Your Own Mail Order Business 0-89384-243-5
Start Your Own Money-Making Computer Business
 0-89384-247-8
Start Your Own Résumé Writing Business 0-89384-249-4
Start Your Own Secretarial Service Business 0-89384-264-8
Start Your Own Temporary Help Agency 0-89384-264-8

Price: U.S. $12.95, Canada $15.95